Studies in engineering education 3
Etudes sur la formation des ingénieurs 3

Titles in this series / Dans cette collection:

1. *Standards for engineering qualifications / Normes de qualification des ingénieurs*
2. *Social sciences and humanities in engineering education / Rôle des sciences sociales et humaines dans la formation des ingénieurs*
3. *The continuing education of engineers / La formation continue des ingénieurs*

e.c.
Studies in engineering education.

# The continuing education of engineers

Proceedings of the
FEANI-Unesco Seminar, Helsinki
21-24 August 1972

# La formation continue des ingénieurs

Compte rendu du
Séminaire FEANI-Unesco, Helsinki
21-24 août 1972

The Unesco Press
Les Presses de l'Unesco 1974

Published by the Unesco Press, 7 Place de Fontenoy, 75700 Paris
Printed by Imprimerie Louis-Jean, Gap

Publié par les Presses de l'Unesco, 7 Place de Fontenoy, 75700 Paris
Imprimerie Louis-Jean, Gap

ISBN 92-3-001141-X

# Preface

The 1972 FEANI-Unesco Seminar on the Continuing Education of Engineers was held at the Helsinki University of Technology, Otaniemi (Finland), from 21 to 24 August 1972. This was the fourth of a series of annual meetings concerned with the education and training of engineers that had been arranged since 1969 by FEANI.

Unesco had originally planned a separate meeting on this topic in 1972, within the framework of its long-term programme on the reform and development of engineering education, but it was decided after mutual consultation that it would be advantageous if only one meeting were held, arranged jointly by the two organizations, so that its scope could extend beyond Europe and its results be more widely disseminated.

The objectives of the seminar were to review the present status and trends of the continuing education of engineers in Europe and around the world, to identify and discuss common problems, and to make recommendations for future action, in particular that to be taken on a regional or international level by FEANI, Unesco and other bodies

To provide source material for discussion at the seminar, a number of papers were prepared by authors eminent in their fields and were distributed prior to the opening of the seminar. Four of these papers were prepared under Unesco auspices, in the form of comprehensive reviews of the status of continuing education of engineers in non-FEANI countries. FEANI arranged for a survey of the situation in FEANI countries, and for the preparation of seven major papers dealing with the different aspects of continuing education that were included in the agenda of the seminar.

A list of the ninety-seven participants from thirty-one countries who attended the seminar appears in Annex 3.

The participants from FEANI countries were nominated by their national member committees for FEANI. Those invited from non-FEANI countries came in their personal capacities.

The seminar was opened in the presence of the Minister of Education of Finland by the Chairman of the Finnish National Committee of FEANI. Addresses were delivered by the Minister of Education, by the representatives of the Director-General of Unesco, and by the President of FEANI.

Two keynote papers, the first on continuing education of engineers, and the second on international co-operation in engineering education, were read during the opening session.

During the Second Plenary Session in the afternoon of 21 August 1972, four papers were presented, describing the status of continuing education in Australia, India, Japan and the Union of Soviet Socialist Republics.

A further paper describing a continuing education programme in the United States of America was then given. This was followed by the presentation of a summary of the continuing education of engineers in FEANI countries.

On Tuesday, 22 August 1972, the main plenary papers were presented and discussed, centred around the themes of the place, the recipients, the content and programmes, the planning, and the means of continuing education.

The seminar split up into working groups on 23 August 1972, to discuss five aspects of continuing education, namely the place, the recipients, the planning and the means of continuing education and the international co-operation in this regard. On the afternoon of 23 August 1972, a number of technical visits were arranged for the participants.

The final session of the seminar was held on 24 August 1972, with the discussion of the reports of the working groups and adoption of the conclusions and recommendations of the seminar.

The present publication contains the two keynote papers, together with summaries of the source papers describing the status of the continuing education of engineers in various countries, and the full texts of the main plenary papers. The publication also includes summaries of the discussions and the reports of the six working groups. Finally, the conclusions and recommendations of the seminar are presented. The addresses delivered at the opening session are reproduced in the annexes. The papers are printed in their original languages and are followed by summaries in English or French. The conclusions and recommendations are given in English, French and German, which were the three working languages of the seminar.

It is hoped that this publication will prove of practical value to those responsible for the planning and implementation of programmes for the continuing education of engineers, to officers of professional engineering societies, and to individual engineers interested in the topic. Those in developing countries who are considering the needs for the creation of systems for continuing education may find it particularly useful.

Thanks are extended by Unesco and FEANI to the Government of Finland and to the Technical University of Helsinki for their hospitality in providing a site and facilities for the seminar, and to the local organizing committee for making all the arrangements necessary for the success of the seminar.

The authors of the different papers are responsible for the material presented and the opinions expressed therein.

# Préface

Le séminaire FEANI-Unesco sur l'éducation permanente des ingénieurs s'est tenu à l'Université de technologie d'Helsinki, Otaniemi (Finlande), du 21 au 24 août 1972. Il constitue la quatrième d'une série de réunions annuelles sur l'éducation et la formation des ingénieurs que la FEANI organise depuis 1969.

L'Unesco avait initialement prévu de convoquer en 1972 une autre réunion sur cette question dans le cadre de son programme à long terme de réforme et de développement de la formation des ingénieurs, mais il a été décidé après des consultations entre l'Unesco et la FEANI, qu'il y aurait intérêt à ne tenir qu'une seule réunion, organisée conjointement, de façon que le champ qu'elle couvrirait déborde le cadre de l'Europe, ce qui permettrait d'en diffuser plus largement les résultats.

Le séminaire avait pour tâche d'examiner la situation actuelle et les tendances de l'éducation permanente des ingénieurs en Europe et dans le monde, de déterminer et d'étudier les problèmes communs et de présenter des recommandations sur l'action à entreprendre, notamment au niveau régional ou international, par la FEANI, l'Unesco et d'autres organismes.

Pour fournir au séminaire une base de discussion, plusieurs documents, établis par des auteurs faisant autorité dans leur domaine, avaient été distribués avant son ouverture. Quatre avaient été préparés sous l'égide de l'Unesco et donnaient un aperçu général de la situation de l'éducation permanente des ingénieurs dans des pays non représentés à la FEANI. La FEANI s'était chargée de faire le point de la situation dans les pays qui lui étaient affiliés et avait élaboré sept documents importants sur les différents aspects de l'éducation permanente figurant à l'ordre du jour du séminaire.

On trouvera à l'annexe 3 la liste des 97 participants, venus de 31 pays, qui ont assisté au séminaire.

Les participants de pays de la FEANI avaient été désignés par leurs comités nationaux pour la FEANI. Ceux des autres pays avaient été invités à titre personnel.

Le séminaire a été ouvert, en présence du ministre de l'éducation de la Finlande, par le président du Comité national finlandais pour la FEANI. Des allocutions ont été prononcées par le ministre de l'éducation, les représentants du Directeur général de l'Unesco et le président de la FEANI.

Deux documents de base, dont l'un traitait de l'éducation permanente des ingénieurs et l'autre de la coopération internationale dans le domaine de la formation des ingénieurs, ont été présentés à la séance d'ouverture.

A la deuxième séance plénière tenue l'après-midi du 21 août 1972, quatre documents, exposant la situation de l'éducation permanente en Australie, en Inde, au Japon et en URSS, ont été présentés.

Un autre document décrivant un programme de formation continue aux Etats-Unis a ensuite été communiqué. Il a été suivi de la présentation d'un résumé des activités liées à l'éducation permanente des ingénieurs dans les pays de la FEANI.

Les principaux documents destinés aux séances plénières ont été présentés le mardi 22 août 1972. Ils portaient sur les thèmes suivants : lieu où est organisée l'éducation permanente, ses bénéficiaires, son contenu et ses programmes, sa planification et les moyens employés pour la dispenser.

Le 23 août 1972, le séminaire s'est scindé en groupes de travail en vue d'examiner cinq aspects de l'éducation permanente ; le lieu où elle est organisée, ses bénéficiaires, sa planification et les moyens employés pour la dispenser, et la coopération internationale dans ce domaine. L'après-midi du même jour, plusieurs visites techniques ont été organisées à l'intention des participants.

A sa séance de clôture, le 24 août 1972, le séminaire a examiné les rapports de ces groupes de travail et adopté ses conclusions et ses recommandations.

La présente publication contient les deux documents de base, le résumé des documents de référence faisant le point de l'éducation permanente des ingénieurs dans divers pays et le texte complet des documents principaux destinés aux séances plénières. Elle contient aussi le résumé des débats et les rapports des six groupes de travail. Enfin, elle présente les conclusions et les recommandations du séminaire. Les allocutions prononcées à la séance d'ouverture sont annexées. Les documents sont reproduits dans leur langue de rédaction et accompagnés de résumés en anglais ou en français. Les conclusions et recommandations sont données dans les trois langues de travail du séminaire – anglais, français et allemand.

On espère que cette publication offrira un intérêt pratique pour les responsables de la planification et de l'application des programmes d'éducation permanente des ingénieurs, pour les dirigeants des associations professionnelles d'ingénieurs et pour les ingénieurs qui s'intéressent personnellement à cette question. Elle pourrait être particulièrement utile aux pays en voie de développement qui étudient actuellement s'il y a lieu de créer des systèmes d'éducation permanente.

L'Unesco et la FEANI tiennent à remercier ici le gouvernement de la Finlande et l'Université technique d'Helsinki d'avoir bien voulu accueillir le séminaire et lui fournir des facilités de travail, ainsi que le Comité d'organisation local, qui a pris toutes dispositions utiles pour en assurer le succès.

# Contents
# Table des matières

I *Introductory lectures read at the Opening Session, 21 August 1972 (morning)*

*Mémoires d'introduction lus à l'ouverture, 21 août 1972 (matin)*

Continuing education for engineers 15
La formation continue des ingénieurs (résumé) 20
Frederick Warner and R. Edgeworth Johnstone

International co-operation in the field of engineering education 22
La coopération internationale dans le domaine de la formation de l'ingénieur (résumé) 26
A. Evstafiev

II *Summaries of papers read during the plenary session, 21 August 1972 (afternoon)*

*Sommaires des documents lus pendant la session plénière, 21 août 1972 (après-midi)*

Continuing education of professional engineers in Australia 31
La formation permanente des ingénieurs en Australie (résumé) 35
A. H. Corbett

The status of continuing education for engineers in India 37
La situation de la formation permanente des ingénieurs en Inde (résumé) 39
S. K. Bose

Continuing education of engineers in Japan 41
La formation permanente des ingénieurs au Japon (résumé) 46
T. Mizushina

Continuing education of engineers in the U.S.S.R. 47
L'éducation permanente des ingénieurs en URSS (résumé) 50
A. I. Bogomolov and M. K. Poltev

The professional development degree in engineering 51
Diplôme de perfectionnement dans le domaine de la formation des ingénieurs (résumé) 55
John P. Klus

Survey of continuing education of engineers in FEANI countries 56
Etude sur la formation permanente des ingénieurs dans les pays de la FEANI (résumé) 70
Osmo Ranta

General discussion 71

III *Main plenary papers presented on 22 August 1972*

*Principaux documents présentés en séance plénière le 22 août 1972*

La formation continue de l'ingénieur en France 75
The continuing education of the engineer (summary) 84
Marc Pelegrin

Demands of society with regard to the continuing education of engineers 85
Les exigences de la société en matière d'éducation permanente des ingénieurs (résumé) 99
Discussion 100
H. Hahn

Content and programmes of continuing education 101
Contenu et programmes de l'éducation permanente (résumé) 108
Discussion 109
M. Jerez Juan, V. Obach, I. Pujol Bayó

Planning of training programmes using job analytical methods 110
La planification des programmes de formation à l'aide des méthodes d'analyse des emplois (résumé) 123
Discussion 124
Kurt E. G. Gestrelius

Modern information services in continuing education 125
Services modernes d'information en matière d'éducation permanente (résumé) 133
Elin Törnudd

Application of technology in continuing education 134
Application de la technologie à l'éducation permanente (résumé) 155
Monroe W. Kriegel

IV *Reports of working groups*

*Rapports des groupes de travail*

The place of continuing education 159

The recipients of continuing education 161

Planning of continuing education 164

The means of continuing education 168

International co-operation in continuing education 170

Discussion in plenary session on 24 August 1972 172

V *Conclusions and recommendations*

*Conclusions et recommandations*

*Beschlüsse und Empfehlungen*

Conclusions and recommendations 175

Conclusions et recommandations 178

Beschlüsse and Empfehlungen 181

*Annexes*

1. Addresses delivered at the Opening Session 187

2. Summaries of additional papers submitted to the seminar 192

3. List of participants 194

4. Members of the local organizing committee 200

# I

Introductory lectures read at
the Opening Session
21 August 1972 (morning)

Mémoires d'introduction
lus à l'ouverture
21 août 1972 (matin)

# Continuing education for engineers

Frederick Warner and
R. Edgeworth Johnstone

## THE NEED

The need for continuing education for engineers arises from a number of factors. First, a formal university or college course leading to an engineering degree is in itself not enough. Engineering is essentially practical, whereas modern academic courses are increasingly slanted towards what is called engineering science. Such courses, however lengthy, have to be supplemented by the exposure of the graduate to practical situations and the necessity for judgements in which science is only one of the factors to be taken into account.

Second, knowledge continually increases and there is a time lag in keeping university courses up to date. Even if it is assumed that a new graduate has been brought up to the frontiers of knowledge in all his studies, the rate of change will make his knowledge obsolete in a short time. In fact, no university course can do more than give students a glimpse of some part of the frontier; it cannot hope to approach the whole periphery.

Third, university engineering studies cannot adequately incorporate the humane disciplines, nor can the inexperienced student profit by them. The small amount which can be done to teach the elements of social relations, industrial organization, economics and ecology at undergraduate level needs to be supplemented after real-life situations have shown the graduate the practical value of more knowledge and understanding.

The need for continuing education in engineering has been widely recognized during the past ten years. In 1964 the American Engineers Council for Professional Development set up a joint advisory committee on continuing-engineering studies. The subject was discussed in Copenhagen at the EUSEC Conference of Engineering Societies of Western Europe and the U.S.A. in 1966. In 1968 the British Institution of Chemical Engineers invited one of the present authors to undertake a study of continuing education in engineering with special reference to the needs of chemical engineers. The study included separate questionnaires circulated to practising chemical engineers and to the firms employing them, discussions with employers regarding their requirements, and visits to educational establishments both inside and outside engineering to study the teaching methods employed in educating mature students. The report on the investigation is available from the Institution of Chemical Engineers, London [1],[1]

1. Figures in brackets refer to the references at the end of this article.

*Table 1.* Continuing-education courses in chemical engineering

| Year | Number of universities and colleges offering continuing education courses | Number of courses | Number of students | | Total |
|---|---|---|---|---|---|
| | | | British | Foreign | |
| 1969 | 1 | 2 | 45 | 2 | 47 |
| 1970 | 2 | 3 | 61 | 40 | 101 |
| 1971 | 5 | 6 | 194 | 28 | 222 |
| 1972 | 10 | 15 | ... | ... | ... |

and its findings need not be repeated here. One of the conclusions was that the methods of teaching are at least as important as the subjects taught, and that the desire for continuing education must be aroused before it can be satisfied. The fact that only about one-third of the questionnaires were returned in each case suggests that interest amongst rank-and-file engineers and their employers was still somewhat lacking.

Since 1969 the Institution of Chemical Engineers has been particularly active in encouraging and sponsoring short continuing-education courses of one or two weeks duration for graduate engineers. Table 1 shows how these have grown during the past four years. The figures refer only to courses sponsored or approved by the Institution of Chemical Engineers. Other British engineering institutions are also active in this field. During 1972 the Institution of Mechanical Engineers offered its members ten continuing education courses in a variety of specialized subjects ranging from 'project management' to 'rapid reading techniques'.

## KNOWLEDGE AND OBSOLESCENCE

Some time ago one of us suggested that engineering knowledge decays logarithmically like a radio-active source and that for a chemical engineer the half-life period is something like five years. E. S. Sellers has proposed a simple mathematical process [2]. Let $S$ represent the total stock of useful knowledge possessed by an engineer; $F$ the fraction of this knowledge which becomes obsolete each year; $R$ the fraction of the engineer's working time devoted to discarding the obsolete and acquiring fresh theoretical knowledge: and $L$ his learning rate. Then:

$$LR = FS + dS/dt$$

($dS/dt$ being the rate of change of useful knowledge).

It is assumed that: (a) $S$ is to remain constant and equal in amount to the stock of knowledge $S_0$ with which the engineer left the university; (b) 5 per cent of this knowledge becomes obsolete every year; (c) the man's rate of learning remains constant and equal to the average rate during a three-year university course. Then the equation becomes:

$$R.S/3 = 0.05\ S_0$$

$$\text{and } R = 0.15$$

The engineer therefore needs to spend 15 per cent of his working time in order to keep his theoretical knowledge up to date.

Engineers are busy people, and it is becoming evident in the United Kingdom that most employers are unwilling to spare them for more than one week per annum for the purpose of attending educational courses. This is only about 2 per cent of an engineer's working time. In the light of the above formula it is obvious that the main task of continuing education must still be carried out by the engineer himself during his work and in his spare time, and that one week's full-time study per annum can make only a minor contribution to his actual store of new knowledge. This suggests that in formal continuing-education courses the major emphasis ought to be on the stimulation of interest and creation of an appetite for up-to-date knowledge that will impel the student on returning to his duties to pursue the search and re-educate himself. In short, the object must be to arouse the desire for fresh knowledge rather than attempt to satisfy it in the short space of a week or so.

## TEACHING METHODS

The report referred to above divides continuing-education courses for engineers into two main types: information courses, the primary object of which is to equip the student with new knowledge; and stimulation courses, the primary object of which is to arouse the interest of the student and impel him to seek out and absorb new knowledge on his own initiative. The distinction is not absolute, since no information course can succeed without some measure of stimulation, and every stimulation course must impart some information; it is a question of relative emphasis. It is perhaps not too fanciful to suggest that this distinction may be associated with the two functions of absorption and elimination. For the mind as for the body, elimination is as important as ingestion; neither mind nor body can continue to absorb new matter unless and until the old is got rid of. Hence, on this view, it is no use trying to pump new information into the minds of engineers unless obsolete ideas and concepts are first removed, and this is perhaps the main function of the stimulation type of course. The removal of obsolete matter automatically creates an appetite for fresh ideas.

The problem of eliminating obsolete methods and attitudes is one which has received little or no attention from educationalists because it is not met with in the initial education of the young. The mind of a first-year college student is usually a *tabula rasa* so far as engineering science is concerned, and the teacher has only to feed in the appropriate knowledge; there are virtually no previous errors and prejudices to be overcome. The mature practising engineer is a different proposition altogether. His initial education and subsequent experience are likely to have impressed upon him some firmly fixed ideas and habits of thought, most of them sound and valuable, but some few obsolete or erroneous. It is these last that have to be brought out into the open and eliminated before new ideas can be effectively implanted.

The stimulation type of course is much more difficult both to design and to teach than the information course. Any young lecturer at a university or college can mug up his textbooks and regurgitate the information in the form of lectures which may be perfectly sound as far as they go; but this type of course can be hideously boring to mature students, especially if it transpires that they themselves know more about the practical aspects of the subject than the lecturer does. One has heard some bitter comments about hastily prepared courses of this kind.

The teacher best able to stimulate the interest of a mature audience is likely to be the experienced engineer who has himself overcome the errors and omissions of his early education, and who can convincingly demonstrate the practical value as well as the intellectual fascination of discarding the old and adopting the new. Such teachers are mostly to be found in industry, and one of the features of continuing education in the United Kingdom has been the extent to which practising engineers have been drawn in to lecture to continuing-education courses held in universities and colleges.

## *QUIS CUSTODIET IPSOS CUSTODES ?*

The mention of universities leads on to a delicate subject which nevertheless ought to be faced; the need for university teachers themselves to be periodically re-educated. Persons who are engaged in teaching are not necessarily learning, and in this respect the university teacher of engineering is in a particularly difficult position. This is because engineering cannot be practised at a university in the sense that chemistry or physics can be practised. Engineering proper is a large-scale co-operative enterprise involving capital, equipment, professional staff and manual workers. A university teacher has mostly to draw upon his past experience of real engineering, though a few of the more senior ones, chiefly professors, manage to keep in touch with industry by acting as consultants. Yet even this is no substitute for the real rough-and-tumble of engineering practice; academic consultants tend to be shown selected problems, and they do not have the responsibility of implementing their recommendations. So the university teacher of engineering stands as much in need of periodically renewing his contact with industry as the industrial engineer does of updating his theoretical knowledge. Once again, the aim must be to stimulate the teacher's interest in practical engineering rather than to instil particular facts.

The need for continuing education in the universities reaches back to the undergraduate course. Returning to the biological simile used earlier, university engineering courses tend to suffer from chronic congestion. New matter is continually pumped in with no corresponding elimination of the old, so that the course becomes gradually distended from four to five to six years. The reasons for this are complex and not unconnected with vested interests that develop within the universities. This is where the professional engineering institutions can play a valuable part, by periodically surveying and re-specifying the kind of education that young engineers currently need. In 1960 one of us had the responsibility of setting up a new chemical-engineering course at a British university. With the co-operation of the Institution of Chemical Engineers a questionnaire was sent to every chemical engineer in the country inquiring which of the subjects commonly taught at universities he actually used in his work. Two-thirds of the questionnaires were returned, and on the basis of these a course was designed which omitted certain subjects [3].The survey was repeated in 1965 in order to reveal trends, and the course content adjusted accordingly [4]. A somewhat similar survey has been carried out by Professor S. P. Hutton and a colleague in respect of mechanical engineering [5].This type of 'consumer research' is one way in which the professional institutions can help the universities to keep up to date.

Where, as in the United Kingdom, professional engineering institutions lay down their own educational requirements for membership, the updating process can also operate in the reverse direction. University teachers are welcomed on the education committees of the engineering institutions, and their influence ensures that the theoretical studies prescribed are in line with the latest developments of engineering science. This two-way interaction between universities and engineering institutions has a revivifying effect on both.

## NEW TECHNIQUES

Our discussion of continuing education has focused attention on the role of engineering societies in identifying the needs of engineers and the response which they are capable of making. For the great majority of engineers, there can be no prospect in the near future of sabbatical years, provided free of financial worry by understanding employers and unencumbered by domestic stress thanks to the willing co-operation of wife and family. The professional societies will make provision of courses, where possible in collaboration with conventional universities, lasting for periods up to one week.

A beginning has been made in investigating the provision of programmed learning for engineers. This is organized to make maximum use of prepared tests, worked examples and questions, with the aid of recorded material for reproduction on a tape recorder or television set. Minimum access to counsellors and tutors is envisaged, but some provision is essential for the student who has not been able to penetrate a particular stage in an argument.

In the United Kingdom a new institution, the Open University, has been set up to provide undergraduate instruction in this way for students who, for one reason or another, have missed the normal educational route giving admission to a university. At present, it is in its second year with 35,000 registered students who receive television and radio broadcasts over the BBC national network. The Open University is now considering the provision of 'post-experience' courses of the type we shall discuss at our seminar. The attraction of this method is that it is almost independent of an employer and relies greatly on the motivation of the student applying to take the course. It makes available to the student an education within a family environment which has already been profoundly changed by the coming of television. We all know the domination over a family time-table exerted by such programmes as *The Forsyte Saga* and the difficulty which engineering societies face in organizing evening meetings in competition with television. The Open University method is capable of providing a new ally for the engineering societies, especially if they can help to see that the back-up facilities such as counselling and tutorial work are provided.

Up to now, programmed learning and audio-visual techniques have been applied most successfully in teaching craft skills and specialized techniques. Their effective application to higher-grade intellectual studies calls for great expertise and, possibly, further research. An early survey conducted by Hoben and Van Ormer for the United States Armed Forces concluded that the effect of audio-visual instruction is liable to be transient unless backed up by solid, old-fashioned reading [6]. Nevertheless, the large number of students that can be reached at one time, compared with the conventional lecture, render these methods important today when the numbers of students threaten to outstrip the supply of experienced teachers.

There are many other open questions in the continuing education of mature engineers; whether short annual courses are preferable to longer courses at less frequent intervals; whether part-time study is preferable to full-time study; the relative merits and demerits of in-company and external courses; and whether external courses are best provided by universities, professional institutions, private consultants, or a combination of these. It is hoped that all of these aspects will be discussed at this conference.

*References*

1. R. Edgeworth Johnstone, *Continuing Education in Engineering,* London, Institution of Chemical Engineers, 1969.
2. E. S. Sellers, discussion, in Johnstone, op. cit., p. 2.
3. R. Edgeworth Johnstone, *Transactions of the Institution of Chemical Engineers,* Vol. 39, p. 263, 1961.
4. R. Edgeworth Johnstone and C. B. Lax, *The Chemical Engineer,* January/February, 1966, p. CE7.
5. J. E. Gerstl and S. P. Hutton, *Engineers: The Anatomy of a Profession,* London, Tavistock Publications, 1966.
6. C.F. Hoban Jr. and E.B. van Ormer, *Pennsylvania State College Technical Report No. SDC 269-17019 (1950).* Sponsored jointly by the Department of the Army and Department of the Navy (NAVEXOS P-977).

# La formation continue des ingénieurs (résumé)

Sir Frederick Warner
Dr. R. Edgeworth Johnstone

Les auteurs se réfèrent à une étude faite pour le compte de l'Institution of Chemical Engineers sur les besoins en éducation permanente des ingénieurs chimistes du Royaume-Uni. L'une des conclusions de cette étude était que les méthodes d'enseignement sont au moins aussi importantes que les matières enseignées, et que le désir de recevoir une éducation permanente doit précéder cette éducation. Ce point a également de l'importance, car le plus souvent, c'est à l'ingénieur lui-même qu'il incombe encore d'assurer pour l'essentiel sa propre éducation permanente, pendant son travail et pendant ses loisirs.

La question de la nécessité pour les professeurs d'université eux-mêmes de se recycler périodiquement se pose aussi, étant donné surtout qu'il leur faut, en outre, reprendre contact avec les techniques de travail de l'industrie. Il est souhaitable que les ingénieurs de l'industrie et les assocations professionnelles d'ingénieurs aient leur mot à dire sur les programmes universitaires et que des professeurs d'université soient membres des comités pour l'éducation des associations d'ingénieurs.

En conclusion, les auteurs examinent l'emploi de nouvelles techniques d'éducation permanente, ainsi que la question de la longueur, de la nature et du lieu des cours d'éducation permanente pour ingénieurs.

# International co-operation in the field of engineering education

A. Evstafiev

When one looks at the history of the engineering profession it is clear that the evolution of the education and training of engineers has been constant, with change following change. In many countries in the nineteenth century there was no formal required education, merely training under the guidance of a senior practising engineer. Then in the present century formal education programmes have been extensively developed and have come to be the dominant factor in supplying the large numbers of engineers required for a modern society.

In the early part of this century the stock of knowledge acquired by an engineer during his education was expected to cover the needs for most of his working life, or at least for that part of it that came before his eventual promotion to a senior managerial or administrative position. Perhaps this occurred after twenty to thirty years of practice of engineering in a technical sense, and during this period the engineer was able to keep up with any advances of technology in his field.

These days have gone. One has only to look at oft-quoted examples of the number of years taken to develop a workable solar battery (about two years) as compared to a workable X-ray tube (about eighteen years, around the year 1900) or the electric motor (about sixty-five years, from 1820 to 1885). Much of the skill and knowledge of an engineer who was educated ten, twenty or thirty years ago is now superseded by major scientific discoveries and their second and higher order effects. Nuclear engineering, the transistor, the computer, the laser, polymers, have all made a vast impact on engineering. The power and complexity of their applications is even now not fully exploited, and yet there will doubtless be many equally striking innovations in years to come. As engineers we must accept the fact that from now on engineers will be called on to work in a dynamic and increasingly complex industrial environment, where changes occur at rapid intervals with interdependent reactions that are unpredictable.

Are we educating engineers in such a way that they will be able to cope with this future world, let alone exploit it? I think that if we are to speak frankly we must admit that we are not doing this, or at least that there is much doubt as to whether the engineer graduating today is well equipped for the tasks that will confront him in five, ten or twenty years hence. There may be even greater doubt with regard to the engineer who graduated ten years ago.

The obsolescence of engineers is now not just a threat, it is a reality. An electronics engineer may have a wealth of experience in vacuum-tube technology, but this now is of little value to him. Many processes in chemical engineering have been superseded by new technologies. Mining engineers and railway engineers may find that they have to change into other fields. On the other hand, interdisciplinary areas are creating many opportunities, and here we need only to quote bio-medical engineering and environmental engineering. Should new engineering disciplines be created to cover these areas, or should their needs be met by engineers training in existing disciplines? These are difficult questions, which can be answered only after much research and study.

To meet these new needs and the rapid developments in existing disciplines, there is an urgent need for reappraisal of our present systems of educating engineers. I believe that we must be prepared to adopt new forms of structures of education, based on systematic research in curricula design and educational methodology, to devise types of education that will be more effective than those used today.

These will inevitably be based on a planned combination of initial education and continuing education, and in this context the subject of this seminar is most timely. Those who have already graduated may consider continuing education as being remedial, updating, or for gaining competence in an adjacent field.

But this is not an acceptable concept for the future, for the graduates of the future. I believe that we should be designing their initial education on the assumption that appropriate forms of continuing education will be available to them throughout their careers. Their initial education should equip them for facing unknown future developments, not just short-term contemporary reality.

These are very difficult concepts and problems for us to face. I believe that collective efforts are required to attack them, bringing together the thinking and experience of engineers in many countries. The Regular Programme of Unesco in engineering education is aimed at facilitating this international collective action, since Unesco is well placed to carry out a catalytic role at the international level.

Perhaps I may briefly describe the history of Unesco's involvement in the education and training of engineers. Until 1960, the activities of Unesco in engineering education were limited, involving mainly the recruitment of individual experts to serve in advisory posts in developing countries. In 1960, however, increased financial resources became available through the United Nations Development Programme's Special Fund. During the next decade over forty large-scale assistance projects were implemented, mostly concerned with establishing or strengthening engineering faculties or similar institutions in developing countries. This has been in general a very successful type of programme, as evidenced by the results achieved at such institutions as: the Middle East Technical University, Ankara; the Indian Institute of Technology, Bombay; the National Polytechnic School, Algiers; the Faculty of Engineering, Nairobi; and the National Polytechnic School, Ecuador. The training of local teaching staff is recognized as the key aspect of this type of assistance. It is noteworthy that in most instances the amount of United Nations assistance was small compared with the expenditure of the host government and with the contributions of bilateral donors, but that its presence and continuity was a key factor in the development of the young institutions.

This UNDP assistance programme is continuing, but with increasing emphasis being placed on programming of assistance on an over-all national basis, rather than through isolated projects. It seems likely that in engineering education there will be more assistance projects for the graduate-level training of engineers in specializations closely tied to local development needs. Such projects also tend to reduce the need to send nationals abroad for long periods of graduate study, and to strengthen research activities.

Assistance projects may play a useful role in the training of certain types of manpower which are in short supply, but they touch only a small number of institutions, and there is a need for other approaches that will reach all institutions and educators. Furthermore, in many countries the quantitative shortages of engineers are now almost overtaken, but there are serious problems concerning the quality of the education provided. The basic scientific content may be outdated and inadequate, the engineering content may be too theoretical in approach to be of any real benefit, the methodology used may stifle initiative and creativity, and the graduate may emerge quite unprepared for the work required from him as a professional engineer.

The remedy for such situations depends finally on the availability of devoted and diligent men committed to improving the efficacity and status of engineering education. There must, however, be some mechanism to encourage discussion and action. This can be provided by a national engineering society, with an appropriate structure of sections or committees concerned with education and training, or by a comparable body in the appropriate government ministry. Alternatively, a single society or group may bring together those interested in engineering education, and the American Society for Engineering Education provides an example of a body of this type. Perhaps it is not too much to hope that similar groups can be formed in other countries, especially in those where engineering educators remain in relative isolation in their own institutions. Appropriate forms of international and regional action can do much to encourage the formation of such groups and to support their work.

In December 1968 Unesco sponsored a major conference on the education and training of engineers, and the recommendations of that conference form the starting point for the Organization's current long-term programme aimed at strengthening international and regional co-operation. This programme foresees, above all, the strengthening of mechanisms for international co-operation. Regional bodies such as FEANI, Pan-American Federation of Engineering Societies (UPADI), and the Federation of Arab Engineers (FAE) have existed for a considerable time and Unesco is co-operating with them, and especially with their activities in engineering education. Support is also being given to the strengthening of other regional bodies concerned with engineering education, in Middle Africa through the Conference on Engineering Education in Middle Africa (CEEMA) and through regional associations now being established in South-East Asia and South Asia.

At the international level, Unesco assisted in 1968 in the formation of the World Federation of Engineering Organizations (WFEO), which took over the role previously played by EUSEC. As might be expected for any new organization of this type, WFEO has faced some initial problems, especially concerning the breadth of its representation and its financial support, but it is now firmly established. Unesco has worked closely with the WFEO Committee on Education and Training and it is hoped that this can evolve into an effective body at the international level, concerned with education and closely linked with the engineering profession around the world.

By means of a series of meetings and consultations, several priority areas for international co-operation have now been identified, around which Unesco programmes are centred and in which work is being carried out, using the above mechanism to the greatest degree possible. Curricula design is perhaps the most important of these areas, in the sense of systematic research on means of framing of curricula that will meet defined objectives and that will be carried out using methodology based on recent knowledge of the learning processes. We hope that our modest Unesco programme will stimulate action in this field, especially with regard to the formation of multidisciplinary research centres or groups that will focus the skills of psychologists, pedagogical specialists, engineers and scientists on the problem of the design of engineering curricula. Unesco would also be pleased to act as a clearing house for the dissemination of information on work being done around the world in this area.

The promotion of co-operation between industry and education is another priority area and one in which there are great disparities between different countries. Some countries are very far advanced, yet others have not been able to achieve any meaningful degree of co-operation, with an absence of communication between engineers in industry and those in education. This is a major problem to be solved, especially in some of the developing countries. The presence of Unesco representatives at this seminar is evidence of the Organization's interest in the area of continuing education, and we propose to follow up the work and recommendations of the seminar with an appropriate programme of activities.

Another area where international action appears desirable is that of the establishment of information services suitable for engineering educators in developing countries.

The role of Unesco in the above fields cannot be one of research and action in depth, as its resources are quite inadequate for this type of task. But it is able to play a role of identifying problems, of stimulating the action of others, and of providing common meeting grounds for those coming from different political or geographical regions. To play this role effectively we need a continuing input of advice from those active in engineering education on a day-to-day basis. Meetings such as this one can help us very much in this regard, as can our contacts with other national, regional, and international groups concerned with engineering education.

As well as the programmes dealing specifically with engineering education, a number of other Unesco programmes are of relevance.

Engineering education is within Unesco the responsibility of the Science Sector, where there are related programmes in science policy, scientific research, engineering sciences and research, and higher technician education.

The Education Sector, which is by far the largest sector in Unesco, has many activities which touch on engineering education, including educational planning, secondary education, continuing education, educational technology, technical education and educational buildings.

Perhaps the most important of the above programmes in the context of engineering education is that for educational planning, where emphasis is placed on the need to assist Member States in the planning and implementation of coherent national systems of education that will be appropriate to the national, social, cultural and economic environment.

The work of the Social Science Sector is also important, in which priority areas are the contribution of social sciences to development and the achievement of Unesco's aims in the sphere of human rights and peace.

In addition to the above, there are several intersectoral Unesco programmes, Man and the Biosphere being a notable example.

Other Specialized Agencies of the United Nations are also working in fields related to engineering education, for example the World Health Organization (WHO), the United Nations Industrial Development Organization (UNIDO), the International Labour Office (ILO), and the United Nations Office of Technical Co-operation (UNOTC). Unesco is in close contact with these bodies.

I would like to emphasize that Unesco's programmes in engineering education are intended to assist all countries ranging from those about to establish their first engineering institution to the most highly developed of the industrialized countries. Lack of communication and lack of co-operation have all too often hindered the flow of information and help between those working in similar fields in different countries, and these are the barriers which organizations such as FEANI and Unesco can break down. I hope that this meeting will adopt concrete recommendations on how we can all work more effectively in developing appropriate continuing-education programmes, as an integral part of the systematic education and training of engineers.

# La coopération internationale dans le domaine de la formation des ingénieurs (résumé)

Dr. A. Evstafiev

Contrairement à la situation qui prévalait au début de ce siècle, où l'on estimait que le stock des connaissances acquises par un ingénieur pendant ses études devait répondre à ses besoins durant la plus grande partie de sa vie active, il est de plus en plus évident qu'en raison des progrès de la science et de la technique, la formation initiale d'un ingénieur doit désormais être considérée comme incomplète et qu'une éducation permanente systématique est nécessaire pour la parfaire ultérieurement. La formation initiale d'un ingénieur devrait en conséquence l'inciter et le préparer à faire face à des situations futures inconnues, et non pas seulement à ses besoins professionnels à court terme.

Convaincue que des efforts collectifs s'imposent pour aborder les problèmes de l'organisation de tels systèmes d'éducation et de formation des ingénieurs, l'Unesco a mis en œuvre un programme à long terme visant à renforcer les mécanismes de la coopération internationale et régionale et à stimuler l'étude de plusieurs domaines prioritaires, dont l'un est l'éducation permanente. Ce programme accompagne et complète les efforts de l'Unesco pour aider ses Etats membres, en particulier les pays en voie de développement, à créer des établissements d'enseignement et de formation des ingénieurs ou a renforcer ceux qui existent déjà. Le manque de communication et de coopération n'a que trop souvent gêné la circulation de l'information et les échanges d'assistance entre personnes travaillant dans des pays différents; mais les efforts d'organisations comme la FEANI et l'Unesco peuvent contribuer à éliminer ces obstacles.

# Summaries of papers read during the plenary session, 21 August 1972 (afternoon)

# Sommaires des documents lus pendant la session plénière, 21 août 1972 (après-midi)

# II

# Continuing education of professional engineers in Australia[1]

A. H. Corbett

Australia is a member of the British Commonwealth with parliamentary government based on universal adult suffrage both in the sovereign states and in the federal sphere, in which the powers of the government are limited by the Constitution originally adopted in 1901.

Under it the states are responsible for education, but the federal government is the principal taxing authority and exerts considerable influence on the way in which the funds it provides for education are spent.

Students proceed at an average age of 18 years to universities at an educational standard known as matriculation, which is roughly uniform throughout Australia. Numerical quotas apply to admission to some faculties. Admission to institutes of technology and colleges of advanced education may be at matriculation or a somewhat lower standard.

Universities are autonomous but rely largely on governments for funds. This restricts their rate of expansion but has not resulted in interference with the internal policies of universities.

Some engineering courses in institutes of technology have been of three years' duration, but the Institution of Engineers, Australia, has given notice that after 1980 only courses of four years' duration will be accepted as a qualification leading to its graduate membership.

Learned societies in Australia have evolved from associations originally modelled on those of the United Kingdom. They are autonomous bodies financed by the annual subscriptions of their members. Some learned societies, including those active in the engineering field, accept responsibility for enhancing the status of the profession and for regulating relationships between their members and the general public. Consequently they are deeply concerned with all matters relating to continuing education.

Definitions of continuing education current in Australia are generally similar to those in use in other English-speaking countries. For example, continuing education may be taken to mean any formal educational activity appropriate for a university graduate other than a programme leading to a diploma or a degree.

In recent discussions on continuing education emphasis is given to a new responsibility of the universities: '. . . the University should be contacting, every year, in some instructing capacity or other, at least as many graduates as it has

1. The full text of the paper may be obtained from the Division of Technological Education and Research, Unesco, Paris.

new graduands. Furthermore, it can be argued that the task of keeping its graduates up to date in a changing and competitive professional environment is even more important to the community than introducing new and professionally immature graduates.'

The generally stated objectives of the continuing education of professional engineers are: the bringing of professional skills up to the standard of current knowledge; the acquisition or improvement of knowledge and skills in specific areas of technology; and, basic or advanced education or training in associated areas such as supervision, management, or administration.

These objectives could be summarized as enabling the professional engineer to increase in stature in his chosen profession, or to qualify for promotion in an associated field such as management.

Australian learned societies have reported that serious consideration should be given to continuing education by young engineers within five years of completing the first degree or diploma. At a time determined by their career prospects they should broaden their interests to include management education or attendance at an administrative staff college.

The trend of future development in the Institution of Engineers is the encouragement of specialist branches in the divisions located in the state capital cities. Each branch is able to serve a group of engineers practising a common discipline in a defined geographic area. The committee of a branch is well placed to evaluate the needs of its members for continuing education, and to advise the division committee on the best means of providing the courses required.

The family responsibilities of professional engineers must be taken into account in estimating their group response to opportunities for continuing education: the period of five to twenty years after graduation is likely to find the professional engineer most receptive.

Two inherent problems in any programme of adult education are the varying ages and the varying backgrounds, both academic and industrial, of students.

The related problem of maintaining a pace which is fast enough to maintain the interest of the bright students but not too fast for the dull students is well known to every academic.

Australian experience has made three contributions to the solution of these problems: the distribution of a preliminary volume of notes which will enable the potentially satisfactory student to revise relevant areas of knowledge and perhaps discourage the attendance of the unsatisfactory student; the arranging of a series of similar courses under administrative arrangements which enable the accumulated experience of the staff to be passed on for future guidance, as in the series of residential Electrical Power Engineering Schools; and the adaptation of methods of instruction to the needs of the students, as in the courses organized by the Department of Foreign Affairs.

The efficiency of a continuing-education programme must be much greater than that of a graduate or undergraduate course, in which the development of the student over a lengthy period may be valuable in itself, and largely independent of the course content. For efficiency the educators must be specialists, and the contribution each is to make to the continuing education programme must be specified carefully.

Students will expect to be issued with carefully prepared printed notes covering the content of the course, and to have adequate visual aids provided and used in lectures.

For the success of an intensive residential course arrangements should be made for at least a portion of the academic staff to reside with the students.

From the students' viewpoint, part-time academic staff may lack even rudimentary teaching skills, and the amount of subject-matter they offer may be excessive in relation to the time allotted. If a number of people contribute to the one course there may be lack of co-ordination between them, resulting in gaps and overlaps in the subject-matter.

The factor of convenience results in non-residential courses being available only in the major cities. Residential courses are generally accommodated in university colleges when undergraduate students are on vacation. There are obvious advantages in locating a residential course in a quiet, relaxing and healthy atmosphere, provided the cost is not excessive and travel facilities are adequate for staff and students.

Effective continuing education of engineers benefits the national economy, the employer, and the employee-student. (In Australia a very large proportion of engineers are employees.) Consequently it is reasonable to expect the government, the employer and the student to contribute to the cost of continuing education, but it must be remembered that each of the three parties will have hidden costs.

In some states student fees are paid into the consolidated revenue of the state, and the university must rely on a matching grant from the state treasury to cover any additional expenditure incurred in relation to continuing education. One director stated that this system worked satisfactorily, and that the state government encouraged his division to extend its services.

In other places it has been necessary for the organizers of courses to set up a financial structure separate from that of the university or college.

For ten years the university of New South Wales Division of Post-graduate Extension Studies has pioneered the application of technology to teaching, and methods first tried in the university have since been widely accepted both in Australia and overseas. Its 'Medical Open Line' is very similar in format to the open-line programmes operated by the various commercial radio stations except that it is available to registered medical practitioners only and the questions are answered by a panel of three or four doctors.

The influence of the division has been extended by the publication of numerous articles describing its work.

Australia's present position is that available resources are being used in systems of continuing education which are less costly than GENESYS, although details of that and other complex systems are well known.

Recently the author was able to initiate negotiations between the University of New South Wales and a group of engineers 2,500 km away in South Australia at Port Pirie for the use of audio and video tapes, which could be exported to any country in which English is the language of instruction. There is no technical reason why any engineering library should not store and lend tapes.

Only in recent years have universities included formal training in the use of information services in undergraduate courses. Consequently the majority of mature and senior engineers require education in the use of information services before the true need for such services can be assessed.

The role of the learned societies is clear: their members must be reminded by a series of short articles in technical journals of the need to become familiar with current methods of disseminating information, and short courses of instruction must be provided.

In the report were summarized programmes of two main types: continuous courses ranging in length from one day to three weeks or three months, some being residential, and part-time courses, usually consisting of evening lectures and discussions repeated at regular intervals. For each programme both administrative and academic details were given, but it has not been possible to describe more than a few examples of the form which continuing education takes in Australia.

This patchwork system has expanded rapidly during the past decade, and is characterized more by energy and enthusiasm than by deliberate over-all planning. At this stage it would be unwise to impose any limitations on its growth, even if it were possible to do so in a free society.

Large industrial enterprises in Australia, with one exception, are owned by overseas financial interests, or depend for their research and development on overseas affiliates. The exception is the Broken Hill Proprietary Company Limited (BHP). While BHP has been a generous supporter of technical education and its senior executives assist learned societies and educational institutions in many ways the company does not appear to have an internal programme of continuing education. Firms affiliated with or owned by overseas interests prefer to send their staffs overseas for experience, which is a form of continuing education, and this policy is approved by the majority of Australians who foster an ambition to see the world, but such experience tends to be concentrated in the early years of an engineer's career, and in his senior years overseas visits are likely to be too short to be of lasting value.

The support given by the electrical power industry to residential schools is detailed in the paper. Further support for continuing education by industry is likely to follow the development of research associations which are being fostered at the present time.

The author believes that industry should not be encouraged to develop internal courses of continuing education in professional fields, but should support the efforts of the learned societies and tertiary educational authorities.

That support by industry up to the present has been adequate is shown by the filling of all places in the courses that have been surveyed in the preparation of this report. The RACI investigation showed that most companies would meet part of the cost and about one-third would meet the whole cost. However, the committee warned that sharing of the cost would be important when the demand for continuing education became widespread. The organizers of a residential school pointed out their success depended upon an initial approach to top management.

## SELF-STUDY

While governments may provide financial support for continuing education administered by the learned societies and offered by educational institutions, the end result depends upon the individual, who will decide whether or not to accept the opportunities that come to his notice, and whether or not to put adequate effort and study time into the course that he elects to attend. For this reason the question of the individual motivation of the professional engineer deserves attention.

Undoubtedly the best kind of individual motivation will result from a sense of professionalism. An outside observer has written that 'the struggle for full professional status for engineering in Australia was part of our struggle for nationhood, maturity and responsibility as a people'. Every member of the profession must be prepared to make a serious investment in the maintenance of that status. For his work to be fully effective it is necessary not only that the engineer constantly improve his skill and widen his knowledge, but also that his employer and the community have complete trust in the integrity, judgement and competence of the profession and of its individual members. If community interests are to be best served each engineer must endeavour to make at all times forthright responsible decisions, and to exercise independence of thought and judgement in his decisions on which his employer and the community depend for guidance.

Representations should be made to the Commonwealth Government for the following administrative changes: (a) expenditure by an individual professional engineer on his continuing education should be exempt from income tax; (b) expenditure by an organization on the continuing education of its employees should be granted double exemption from company tax.

## IDEAL DEVELOPMENT

Continuing education in future should result from a partnership between governments, learned societies, educational institutions, employers and employee-students.

The examples described in the paper show that the major elements of such a partnership already exist in the Australian community. The learned societies have the opportunity and the responsibility to act as catalysts to promote further progress.

The Institution of Engineers, Australia, has set up a National Committee on Engineering Education, responsible to its council through the Science and Practice of Engineering Committee. The responsibilities of the National Committee will include continuing education.

# La formation permanente des ingénieurs en Australie (résumé)

A. H. Corbett

L'auteur étudie le système australien de formation des ingénieurs et d'enseignement professionnel et indique la situation actuelle des activités d'éducation permanente.

Ces activités sont assurées par les universités et les organisations professionnelles d'ingénieurs. L'auteur décrit, à titre d'exemple, les cours de brève durée, avec hébergement sur place, qui sont organisés à l'intention des ingénieurs en électricité. On utilise de plus en plus souvent le magnétophone et le magnétoscope pour la formation permanente des ingénieurs, notamment quand ceux-ci résident loin des grandes villes.

L'auteur souligne que l'éducation permanente va sans doute se développer rapidement et, entre autres recommandations, préconise d'encourager ce mouvement au moyen d'incitations fiscales. En conclusion, il insiste sur le fait que la formation permanente doit faire l'objet d'une collaboration entre le gouvernement, les organisations professionnelles d'ingénieurs, les établissements d'enseignement, les employeurs et les ingénieurs eux-mêmes.

# The status of continuing education for engineers in India[1]

S. K. Bose

The stockpiling of scientific and technological knowledge demands the continuous addition of new knowledge to the old and the placing of the old knowledge into new relationships. The quality and quantity of this increasing stock of knowledge determines the nature and extent of its acquisition. Some form of further learning and training after the initial formal engineering course seems to be an inevitable requirement today. There is, moreover, a proliferation of scientific knowledge in the border areas resulting in the growth of the interdisciplinary phenomena. On the other hand, expeditious utilization of new knowledge demands that engineers should acquire it within a short period, in order to fully benefit from it. What the changing situation really demands is a continuous review of the needs with a view to introducing a suitable reorganization of the structure of engineering education vis-à-vis general education of which it forms a part.

Continuing education aims at specific enhancement of competence of individuals as engineers rather than the attainment of additional academic qualifications. The process of enhancement of individual competence is so planned that it keeps the individual abreast through educational programmes designed to provide the necessary background to match the required competence with the diversity needed.

These programmes should provide independent, self-sustaining courses of learning which are specifically designed and developed, normally requiring a much greater effort than conventional courses. Such courses constitute an important part of the over-all system of education in a developing country. They not only make a contribution to productivity by improving knowledge and skill of those in employment, but also help to reduce the pressure on facilities for full-time education.

The term 'engineers' used in this report embraces the products of all recognized engineering institutions offering basic engineering courses and diplomas in engineering, i.e. post high-school courses of five years' duration as well as of three years' duration.

The report makes a survey of engineering education in India, indicating the enrolment of students, the annual output, the imbalance in the growth of engineering manpower in relation to supply and demand in various main branches of engineering. It also describes the low ratio of production of diploma-holders

1. The full text of the paper may be obtained from the Division of Technical Education and Research, Unesco, Paris.

(technicians) to engineer executives. To the latter, continuing education must provide remedial measures for updating in their own specialized fields and for acquiring competence in allied fields. The problem faced by India needs several corrective actions. In 1955 the number of engineers was 79,000 and in 1966, 332,000. It is expected to reach 465,000 by 1975 and 564,000 by 1978. In addition, a significant number of positions, mainly in the production sections, are held by 'practicals' trained within plants or works, accounting for approximately 25 per cent of all engineers in employment. Different courses of continuing education are indicated in the report for those qualified in basic engineering courses, diploma courses and those without formal technical education (practicals). There is also a significant imbalance in the availability of engineers in different branches of engineering. Moreover, the present ratio of middle-level technicians to engineer executives is low, yet unemployment among the former is higher than among the latter.

In India the programmes of continuing education are supervised and co-ordinated by the same academic committee responsible for organizing the advanced engineering courses in many institutions. The scope of this work has been enlarged but has not reached a stage demanding the constitution of a separate department of continuing of a separate department of continuing education. This has a disadvantage. Continuing education cannot conform to the general pattern of organization of other departments. Its function has to be of the nature of a multidisciplinary department in which the head of the institution and other authorities of the institution must take a special interest tor its working and development because its areas of operation often extend beyond the confines of the university or institution and proper liaison must exist between it and the community it serves.

Where not established as a separate unit, the multidisciplinary nature of these activities make it imperative that a separate unit should be brought into existence. In one or two universities a separate unit has been established. To ensure the proper functioning of the unit, it has been found necessary to set up an advisory committee with the vice-chancellor or equivalent as its chairman, with representatives of both the academic community as well as professional bodies within the area served by the institution. The important thing is to ensure that the programmes developed make an impact on the neighbouring community who should feel a sense of involvement both in the planning and the conduct of the programme of continuing education for technical personnel.

The programmes of continuing education require separate and additional funds for their organization and sustenance. In institutions where these courses are run the sources of financing are grants from the university and the government as well as the income from fees. In big cities, the programmes can be sustained through income derived from fees paid by students. In their desire to improve their skills, the students are prepared to pay for this educational opportunity. Moreover, charging fees for students has its merits. The quantum of fees to be charged is also dependent upon the payment to be made to the teachers, both internal and external, for their additional work.

There is hardly any aspect of national life which is not directly dependent on the extent and quality of education. The need for adequate provision for continuing education has now been realized. The close relationship between the education of engineers which must incorporate continuing education for them and development demands that the programme of continuing education be a national concern, and be financed accordingly by the same body which funds technical education.

In India, a centrally sponsored scheme for quality improvement of the teaching staff of engineering institutions by means of continuing-education programmes has been a success. Details of the programmes are given. However, much remains to be done for professional engineers. Several institutions organized full-time post-basic diploma courses of one year's duration to meet industrial needs. All, however, did not receive adequate response in the form of sponsored candidates, and it has been found necessary to offer these courses on a part-time basis at convenient centres. More work must be done on the provision of facilities for diploma-holders. There is also a growing need to identify areas in which courses for professional engineers have become necessary and to organize in-service training programmes for them. This is also true of the 'practicals'. Professional bodies are working to this end. There are, however, many organizational problems to be overcome.

When offered along with the regular academic programme, programmes of continuing education have created problems. Incentives to teachers for active participation in the form of reduced teaching load, convenience of travel, and monetary benefits have been suggested. Industrial participation in providing training has so far been encouraging. Many industries are offering better financial benefits to chosen students than those offered by the government stipendary training schemes for the training of engineers after they leave the institutions. Many industries have not, however, established professional development committees to review and initiate programmes of continuing education for those already working with them. Professional bodies, twenty-two in number, offer graduateship examinations twice a year and are now organizing refresher courses for practising engineers in conjunction with teaching institutions. Their increasing interest in organizing continuous-education programmes is encouraging.

The present trend is towards consolidation of university-level institutions rather than quantitative expansion and towards giving adequate autonomy to selected polytechnics for the training of technicians and the initiation of new experimental projects in technician education. One of the proposals under consideration is to create State Councils of Technical Awards to replace the existing governmental control of the technician training in polytechnics.

# La situation de la formation permanente des ingénieurs en Inde (résumé)

Professeur S. K. Bose

L'auteur décrit le système de formation des ingénieurs en Inde, en donnant le nombre total des inscriptions dans les établissements d'enseignement et celui des diplômés, ce dernier chiffre étant mis en relation avec les besoins en personnel du pays. Il indique aussi le nombre des ingénieurs "autodidactes". Il existe actuellement un déséquilibre notable entre les effectifs d'ingénieurs de diverses spécialités, ce qui se traduit par une pénurie de personnel dans certaines branches et par du chômage dans d'autres. Dans de nombreux domaines, on manque aussi de techniciens compétents.

L'Inde a appliqué avec succès certains programmes de formation permanente, notamment un programme de perfectionnement des professeurs des sciences de l'ingénieur. Toutefois, il s'agit surtout d'efforts isolés et aucun système d'ensemble n'a encore été mis en place. Dans les facultés d'ingénieurs, des difficultés ont parfois été soulevées par la rémunération des enseignants, par l'absence de structures adaptées à la formation continue, et par la question du financement. L'intérêt croissant que les organisations professionnelles et les milieux industriels portent à la formation permanente est un élément encourageant, mais il reste encore beaucoup à faire, notamment du côté des entreprises.

# Continuing education of engineers in Japan[1]

T. Mizushina

## INTRODUCTION

Continuing education has not yet been systematized in Japan, and the facilities and courses that exist have been designed on an *ad hoc* basis to meet the explicit and implicit demands of society, hence the majority of engineers obtain knowledge of new developments in their professional areas mainly by self-teaching with the aid of books.

Because of the rapid development in every area of engineering in recent years, the knowledge obtained in schools becomes inadequate within a few years unless a programme of continuing revision is practised. In addition, the demands of engineers for knowledge and technology which have been developed recently, such as information technology, emphasize the necessity for continuing education. For example, since it is difficult to establish in universities the number of departments of computer science necessary to produce a large number of graduates within a few years, re-adaptation of engineers of different specialities through continuing education in computer science is the only way to meet the tremendous demand. Accordingly, the curricula in engineering schools put more weight on engineering sciences, rather than practical aspects, so as to give a student a basis for his career as an engineer. On the other hand, the importance of continuing education after graduation is emphasized.

To understand the status of continuing education of engineers in Japan, it should be noted that there are several unique features.

First, a Japanese engineer is quite often employed by a particular company for his lifetime.

Second, there is no registration system for engineers. Accordingly, there is no definite distinction between engineers and technicians.

Third, besides the ordinary Ph.D. course in universities, one may obtain the doctor's degree by submitting a thesis. Accordingly, even a technician can be regarded as an engineer provided that he has continued to educate himself to fill such a position. Many individuals who are chief engineers in firms or full professors of colleges have only the educational background of technological secondary schools. And anyone—even a man who received only the nine years

1. The full text of the paper is available from the Division of Technological Education and Research, Unesco, Paris.

compulsory education—who educates himself so as to reach a certain level of knowledge can apply for the doctor's degree. In 1970, 51 course graduates and 70 non-graduates, and in 1971, 43 and 81 respectively, received doctor's degrees of engineering from Kyoto University. These are strong motive forces which make some Japanese continue their education after graduation.

Fourth, Japanese technicians have a high level of ability and thus employers pay attention to their continuing education in order to bring their knowledge and techniques up to date. Moreover, the period of rapid industrialization after the Second World War when Japan needed a rapid increase in the number of engineers, initially needing quantity rather than quality, has produced a situation in which she must now improve the quality of her engineers. For this purpose also, the continuing education of engineers has recently become important.

The first part of the paper explains the present educational 6-3-3-4-2-3 system and the recent expansion in engineers' education of Japan.

Then a summary is given of a statistical study of the present situation of continuing education of engineers in industry, which was carried out by the Science and Technology Agency of the Japanese Government through contact with the employers.

Next, several examples of current courses of continuing education are studied.

Finally, the section relating to continuing education in an advisory report of the Central Council for Education in the Ministry of Education, which was presented to the Japanese Government in 1970, is extracted since it is to some extent relevant to the future.

## PRESENT EDUCATIONAL SYSTEM

To appreciate the present situation and problems of continuing education in Japan, it is necessary to know about the present system of education. Before the end of the Second World War, a 'multi-track' system was adopted in Japan, but it was then changed to a 'single-track' 6-3-3 system.

From all the different types of educational institutions, those relevant to engineering education are as follows.

*Upper secondary schools (Kōtōgakkō = high school).* Following nine years' compulsory schooling, there are three years of upper secondary school. The upper secondary courses are classified into general and vocational courses, which are further classified into several categories, including technical courses. There are both full-time and part-time schools in technical courses. The number of technical high schools was 715 in 1970.

*Universities (Daigaku).* The requirement for admission is the completion of upper secondary school or its equivalent. Courses for bachelor of engineering degrees extend over four years. There are graduate schools in most universities. The lengths of the master's and doctor's degree courses are respectively two years and five years beyond the bachelor degree. The number of universities offering engineering education was 105 in 1969.

*Junior colleges (Tanki-daigaku).* These schools provide upper secondary school graduates with two- or three-year courses. However, the majority of such courses are for girls, and relatively few are for engineering education. In 1970 there were forty junior colleges with engineering courses.

*Technical colleges (Kōgyō-kōtō-semmon-gakkō).* These are a combination of a technical upper secondary course and junior college course, and offer five-year continuous programmes for training technicians. Technical colleges started in 1962, and in May 1971 there were sixty-three such colleges.

*Miscellaneous schools (Kakushu-gakkō).* In addition to the regular schools, there are a variety of educational establishments known as *Kakushu-gakkō*. Most of them are privately operated and provide their students with vocational and practical courses. The length of courses vary; most are one year but some may be as short as three months. Within this category are schools for computer science and some of the institutes for continuing education, both of which play a role in the continuing education of engineers.

In recent years, the capacity of the schools has been expanded to absorb the greater number of children born after the Second World War. In order to meet the needs of the industrialization of Japan, there has been a considerable expansion of facilities for engineering education. As a result, the percentage of engineering students to all university students reached 20 per cent in 1969, and about 5.5 per cent of boys of the relevant age group were taking engineering courses in university.

## STATISTICAL STUDY OF CONTINUING EDUCATION OF ENGINEERS IN INDUSTRY

The Science and Technology Agency of the Japanese Government reported the results of the Statistical Study of Continuing Education of Scientists and Engineers in Industry in March 1970. The aim of the study was to determine the present situation of continuing education in industry in order to obtain the basic information for making a governmental policy for the continuing education of scientists and engineers in industry.

The study was executed by correspondence with those industries whose capital exceeded 1,000 million yen. The personnel studied were classified into two categories, namely researchers and engineers.

### *General aspects*

About 65 per cent of companies carried out a programme of continuing education in 1968. Roughly 70 per cent of the programmes were inside the company, and 30 per cent outside the company. Among the facilities in Japan outside the company, universities and research institutes are not often used. The reason for this seems to be the fact that these facilities cannot accept many additional persons, or that they cannot meet the needs of the industries concerned.

The period in which the employees take continuing education courses is mostly 3-10 years after their graduation. The engineers are probably expected to demonstrate their abilities fully in their work during this period.

The length of time involved for continuing education courses is mostly less than one month (40-150 hrs).

*Tendency to increase involvement in continuing education*

More than 40 per cent of companies indicated a tendency to become increasingly involved in courses of continuing education. The number of employees taking continuing-education courses has increased rapidly in recent years. The divisions in which the need for continuing education has increased are those of research and development and of production engineering, where tasks have become more complex and more advanced with recent developments in science and technology.

*Improvement of the system of continuing education*

As for the difficulties of industries in carrying out continuing education, it was mentioned that they are too busy to spare the time for continuing education, that good facilities for education are scarce, and that appropriate lecturers can hardly be found. To solve these problems industries demand that efforts be made to establish special institutes for continuing education and to use effectively universities and governmental research institutes. Most industries prefer to press for education outside the company rather than that inside the company, because of their limitations in educational facilities, lecturers and subjects.

## EXAMPLES OF CONTINUING EDUCATION

*Courses sponsored by the Japanese Society for Engineering Education.* The Japanese Society for Engineering Education has a Committee of Continuing Education which has a subcommittee consisting of seven professors of different universities who have since 1968 been organizing continuing education courses.

*Courses in institutes for continuing education.* There are several institutes which have been established in Japan for the purpose of continuing education. Among them, the Advanced Technical Training Institute of Osaka may be the first and is a typical and most well-organized facility. Another organization offering various continuing education courses that differ from those of the Advanced Technical Training Institute of Osaka is the Union of Japanese Scientists and Engineers in Tokyo. Besides work in publications and in making computer software, the Union is operating several courses for continuing education.

*Seminars sponsored by engineers' societies.* Seminars sponsored by the engineering societies may now be the main force of continuing education in Japan conducted outside companies, in the number of participants, their frequency and their variety.

*Courses in industrial companies.* Among the companies studied, a manufacturer of electrical and industrial machinery (capital, 122 billion yen; number of employees, 93,700) has established the best developed internal system of continuing education, the Board of Directors having adopted in 1968 a policy to promote an educational system.

*Continuing education in university.* Virtually no continuing-education courses are conducted in Japanese universities, although many professors are invited to lecture in courses outside the universities. However, the system of research fellows and non-credit students may correspond to some extent with the objectives of continuing education. For example, in 1971 in the Faculty of Engineering of Kyoto University, the second biggest faculty of engineering in Japanese governmental universities, having 147 full professors, there were 211 research fellows (university graduates), 11 research aids (high-school graduates), and 65 non-credit students sent from industries.

*Courses on radio and television.* Several trials of courses and open university on radio and television are being made.

*Courses for overseas students.* Several seminars for overseas students have been conducted in Japan as the result of co-operation with Unesco. These seminars are part of the Technical Cooperation Scheme for Developing Countries of the Japanese Government, and were sponsored by Unesco and the Overseas Technical Cooperation Agency (OTCA) of Japan.

*Other programmes.* As examples, seminars sponsored by the Information Technology Seminar Centre, Nippon Electronics Practice Institution and the Industrial Daily News Company are studied.

## ADVISORY REPORT OF THE CENTRAL COUNCIL FOR EDUCATION IN THE MINISTRY OF EDUCATION

After about two years of discussion, the Central Council for Education presented an advisory report on the reconstitution of the educational system to the Japanese Government in 1970. The section relating to the continuing education facilities is as follows:

Chapter 2. Basic Concept of the Reconstitution of Higher Education System
(i) Diversification of higher education system.

It is recommended that the higher education facilities be classified by the difference in qualification for admittance and the number of years necessary for completion and that the several types of courses be arranged for each objective and character of education as follows. At the same time such educational system must be prepared so that the students can move to another category of facility or another type of course when it is necessary.

(1) Category A of higher education facility – University
(2) Category B of higher education facility – Junior College
(3) Category C of higher education facility – Professional College
(4) *Category D of higher education facility* – Graduate School
(5) Category E of higher education facility – Research institution

The fourth category is for offering 2 to 3 years' advanced courses in certain special areas to the graduates of universities or to men who have obtained the corresponding knowledge by self-taught or practical experience, and for the separate units of *continuing education course* to people working in society. These courses will be at the level of those in the master course in the present educational system.

This 'Graduate School' may be installed within the 'University' to offer a continuous education as in the present system.

## CONCLUSION

The divergence of the needs for continuing education of industries may well continue, as a solution common to all kinds of industry may not be reached in spite of the present eager discussions about it. Accordingly, such a variety of patterns and subjects of continuing education as mentioned in this paper will continue to exist or even increase. Therefore, systematically arranged information about the continuing-education courses available should be given to all industrial companies and individual employees. Furthermore, it may be desirable that systems for advising individuals to select courses and in setting up over-all programmes to meet their objectives will be established in engineers' societies and in industrial companies.

# La formation permanente des ingénieurs au Japon (résumé)

Professeur Tokuro Mizushina

L'auteur expose en détail les différents modes de formation des ingénieurs utilisés au Japon. On relève notamment qu'il n'existe pas de système d'immatriculation pour les ingénieurs, qu'il n'est pas rare qu'un ingénieur travaille dans la même entreprise pendant toute sa carrière, et qu'on peut obtenir un doctorat soit au terme d'études universitaires, soit en soutenant une thèse concernant des recherches d'un intérêt exceptionnel. Dans ce dernier cas, il arrive que l'intéressé n'ait pas suivi d'enseignement de type traditionnel au-delà du second degré, tout en ayant atteint un très haut niveau professionnel.

En 1969, 109 universités et 40 collèges universitaires assuraient un enseignement de sciences de l'ingénieur d'une durée respective de quatre et de deux ou trois ans. En mars 1970, le gouvernement a fait effectuer une enquête sur la formation permanente des scientifiques et des ingénieurs au Japon. Des renseignements très intéressants ont été recueillis, notamment sur l'activité des milieux industriels dans ce domaine.

Le document donne quelques exemples de programmes de formation permanente dont ceux de la Société japonaise pour l'éducation des ingénieurs, de l'Institut de formation technique supérieure d'Osaka, de l'Union des scientifiques et ingénieurs japonais, et de diverses associations professionnelles d'ingénieurs et entreprises industrielles.

L'auteur conclut qu'au Japon la formation permanente des ingénieurs continuera d'être assurée par des moyens variés et qu'il faudra conseiller les intéressés pour les aider à choisir les cours et à élaborer les programmes qui répondent aux objectifs visés.

# Continuing education of engineers in the U.S.S.R.[1]

A. I. Bogomolov and M. K. Poltev

As the industry of the U.S.S.R. has developed, the need has grown to expand and improve the national system of continuing education for engineers. It has become especially necessary to give engineering personnel knowledge in such fields as economics, and in new areas of exploration in science or technology, or in cases where there is an urgent need for expansion of existing branches of the national economy.

Basic engineering education in the U.S.S.R. has a duration of five to six years. This includes systematic practical training that includes a variety of work ranging from that of a workman to that of an engineer. At the end of the course the student has to prepare a diploma project for presentation to the State Examination Board. The basic engineering qualification may also be obtained by part-time studies. Those interested in a career in research may continue their studies for a further two to four years and obtain the academic degree of Candidate of Science. Beyond that degree, completion of major scientific work in science or technology can finally lead to the award of a Doctor of Science degree.

There are unified standards in the U.S.S.R. for the structure and instructional procedures of all categories of continuing education for engineers. It is under the over-all co-ordination of the Ministry of Higher and Specialized Secondary Education, regardless of whether it is carried out in educational establishments, special training institutes, ministries, or in individual industrial enterprises. Three distinct types of continuing education are recognized, namely additional education, retraining and advanced training.

## ADDITIONAL EDUCATION

Additional education is designed to give engineers a more profound knowledge in a specific field. An example of additional education is an evening course in mathematics of a duration of three years and nine months, with admission competitive and on the recommendation of the industrial enterprise employing the engineer. Of the total time spent amounting to 199 weeks, 133 weeks (or 1,922 hours) are spent on theoretical studies, 27 weeks on course credit tests and

1. The full text of the paper is available from the Division of Technological Education and Research, Unesco, Paris.

examinations, 4 weeks on final State examinations, and 35 weeks on vacations. Close to 15 per cent of the time is devoted to alternative disciplines such as additional topology, additional logic, celestial mechanics or gas dynamics.

Another example is a two-year full-time course in applied machanics and automatic control systems, emphasizing computer methods. At least two years of related work in industry is a prerequisite for the course and at its end the engineer is awarded a second diploma.

Managerial courses of three to six months' duration are another example. These are for factory managers or engineers who are in senior managerial positions and yet not over 40 years of age. For instance, in the industrial planning faculty of the Moscow Institute of National Economy the subjects covered include computers and processing of economic data, statistics, scientific fundamentals of management, legal aspects of management and economics, planning of the national economy, accounting of various industries, etc. The aim of the course is to adequately fill the gap in the students' knowledge arising from his promotion to a managerial post.

## RETRAINING

This is further education to give an engineer competence in a field other than that of his basic education, which will open up a new avenue of work for him. For example, graduate engineers are retrained in artistic design in a three-and-a-half-year part-time course. On completion of the course they are awarded a second diploma and are then normally engaged as industrial designers of new manufactured products.

One example of a specialized educational establishment solely engaged in retraining is the Central Institute of Patent Science. This institute accepts graduates for either a six-month full-time course or a two-year course for those who study through evening courses or correspondence. The full-time students receive stipends from their employing organizations and the part-time students receive an additional ten-day study leave each year and various other privileges concerned with travel and subsistence expenses.

The curricula are elaborated by the institute in co-ordination with the Committee for Inventions and Discoveries and with the approval of the Ministry of Higher and Specialized Secondary Education. A total of 1,580 hours are spent in class, of which 40 per cent is allotted to lectures, 50 per cent to seminars, and 10 per cent to consultations and examinations. Not only are U.S.S.R. patent law and procedures covered, but international economic relations and patent practice are also studied in depth, especially with regard to protection and licensing arrangements.

## ADVANCED TRAINING

Advanced training of engineering personnel has assumed large proportions in the U.S.S.R., with a ramified network of training establishments in every branch of the national economy. These establishments may take a number of different forms, namely: advanced training facilities at higher educational establishments; advanced training institutes serving one or several industries, referred to as departmental or interdepartmental respectively; advanced training courses set up within ministries; advanced training courses set up at individual enterprises or educational establishments.

The above network is designed to provide training units for all enterprises, such that all engineering personnel can receive training at specified intervals of every four or five years. At present, facilities exist for offering schooling to up to 1 million specialists per year. The supervision of the methodology used is entrusted to the Ministry of Higher and Specialized Secondary Education of the U.S.S.R., which has established a Methods Science Centre to be responsible for research and co-ordination of all continuing education. The individual industrial ministries and departments in need of continuing education for their personnel are entitled to come up with proposals for the establishment of appropriate facilities. Within an individual ministry, an advanced training institute may be the apex of a network of advanced training units located in individual enterprises or elsewhere, and this institute will then be responsible for curricula used, teaching materials, and the collection and dissemination of information.

The length of the courses may vary, but normally will be up to two months for full-time or six months for part-time courses. Some, however, may be equivalent to one month of full-time study and in areas such as computation they may be as long as four months' full-time or ten months' part-time study. The students is paid his normal salary and various allowances for travel and subsistence.

Once established, an advanced training centre grows into a well appointed educational institution, able to provide deeply penetrating courses and give the student a systematic review of all recent developments in his field.

As an example, the Moscow Institute of Steels and Alloys has a two-month full-time course for directors, chief engineers and shop superintendents of enterprises in non-ferrous metallurgy. Of a total of 390 hours, 60 hours are specifically devoted to non-ferrous metallurgy, 180 hours to economic and management topics, 60 hours to electives, and 90 hours of seminars and project.

Similar types of programmes are offered at the Moscow Electronics Institute in such topics as 'Production of Integral Microcircuits' and 'Fundamentals of Microelectronics'.

At all such institutes the curricula may be tailored to suit the needs of specific categories of students. Thus at the Advanced Training Institute of the Ministry of Light Industry of the U.S.S.R. there are separate programmes for directors of enterprises, chief engineers, chief mechanical engineers, and specialists in financial management, with different proportions of time being spent on hardware and technology, economics, and labour and management. In every case, on completion of the course each student is expected to carry out a project that is related to his work.

The new system of advanced training outlined above places a heavy responsibility on those involved in its administration and teaching, especially to ensure that high-quality schooling is provided. The problem of raising the efficiency of instruction under these new conditions is most important, and much work is now being done in the U.S.S.R. in this regard.

# L'éducation permanente des ingénieurs en URSS (résumé)

A.I. Bogomolov et M.K. Poltev

Ce document contient un exposé complet du système d'éducation et de formation des ingénieurs en URSS, suivi d'une description des trois sortes de programmes de formation permanente: éducation complémentaire, recyclage, et formation avancée; des exemples de programmes typiques sont donnés. Le système repose sur le principe de la création d'un réseau d'unités de formation pour toutes des entreprises d'une branche industrielle donnée, généralement avec un ou plusieurs instituts de formation avancée pour assurer la direction de l'ensemble, et souvent avec la collaboration d'autres branches industrielles. Le Ministère de l'enseignement supérieur et de l'enseignement secondaire spécialisé de l'URSS a été chargé de la supervision générale, en ce qui concerne les méthodes d'enseignement de l'ensemble du système de formation avancée du pays. Il a créé un Centre de méthodologie qui s'occupe des méthodes à employer pour l'application des programmes, des échanges d'informations et du travail d'évaluation.

# The professional development degree in engineering[1]

John P. Klus

Definitions of continuing education often vary within the academic community and in industry. Many think of residence or evening credit courses that lead to traditional advanced degrees. Quite often we cite continuing education as embracing all those programmes, such as institutes, workshops and conferences, of one or several days' length, and short courses of one or two weeks' duration, which have evolved as service to industry yet would not be recognized for regular degree programmes.

When we consider technology development, the day-by-day pressures on the engineer at work, job mobility, and wide geographic distribution of both engineers and educational resources, a new approach to continuing education seems not only apparent, but necessary. Residence requirements detour many from the path to traditional degrees. Prerequisites or fixed curricula inhibit others. To some it has been helpful to take short courses, attend seminars, or engage in other study formats selected at random, but this is not an efficient solution to the average engineer's problem.

Wisconsin's professional development (P.D.) degree programme offers an opportunity to satisfy the engineer's need for structure and goal. It is a new post-graduate degree pathway for adult working engineers, not a degree between the bachelor's and the master's, nor a degree to replace any other, but one that can uniquely fit into the continuing education life, development, and the means of an average practising, on-the-job engineer. It is a degree programme very specifically oriented to engineers who have full-time jobs, those without time to return to full-time resident classroom instruction. It allows engineers to remain essentially fully active in their daily work, to blend their practical experience with study, and yet be enrolled in an individually structured course of study with an achievement goal in sight—the P.D. degree—and it permits the full resources of the University of Wisconsin, and those of other universities as well, to bear upon any individual's programme of studies.

The particularly significant aspect of this P.D. degree programme is that hours of credit are available to the candidate from all the continuing education types of programmes that extensions have so widely developed: intensive short courses, institutes, evening courses, seminars and workshops, correspondence studies, guided individual project study, and so on.

1. The full text of the paper may be obtained from the Division of Technological Education and Research, Unesco, Paris.

'Continuing education' and 'extensions' have the important role of providing educational opportunities through a variety of formats. But to be of most value, standards must be found and adopted that are acceptable to both the academic and industrial worlds. The professional development degree of the University of Wisconsin establishes such standards and sets a pattern of education which has attained academic and industrial recognition equally.

## GUIDELINES AND REQUIREMENTS

The programme accepts any candidate who can at least qualify as having earned a bachelor's degree in engineering or a science degree from an equally accredited programme; for example, physics, chemistry, mathematics. Previous college grade-point average does not influence acceptance as a candidate. There is no residence requirement. There are no fees for registration. It should also be noted and recognized that the P.D. degree programme will be valuable to the person with an M.S. or even a Ph.D. degree as well as to the holder of a bachelor's degree.

To earn the degree a candidate must compile a minimum of 1,200 hours of credit (study) in his programme. This equates in rough terms to twenty-five undergraduate credits. Why 1,200 hours? Psychologists and others were questioned as to what commitment to study by the average full-time practising engineer could be expected. The answer was that about six hours per week would be reasonable. In one year this becomes 300 hours and in four years would total 1,200 hours, a span of time that could enable most to finish. Of that total, it is required that most of the earned hours of credit (at least 1,000) be evaluated by examination or written report. Not more than 200 hours from programme participation may be earned without examination. The latter does allow considerable engagement in worth-while seminars and institutes.

There are two time-limitations imposed on a candidate. First, the P.D. degree cannot be granted within three years of the data of his most recently earned degree. We wish to ensure that those earning this degree will establish some blend of experience and education. Second, a limit of seven years is allowed to compile the hours of credit required to qualify for the degree. If one's programme continues beyond that, hours earned in the beginning year will be in jeopardy. This requirement is not uncommon to most technical educational programmes, the master's and doctorate included.

The over-all guidelines for any individual's P.D. degree programme emphasize very strongly that a balance of time be planned between different elements of study. These elements we have characterized as:

1. *Technical updating* – reviews, refreshers (not more than 20 per cent, 240 hours, of total). With this updating and refreshing built into the P.D. degree programme students may more willingly pursue the refresher courses, when they perceive their need, before starting more advanced studies.
2. *Technical advancement upgrading* – new technology developments (at least 30 per cent, 360 hours). It is only logical, since this is a P.D. degree 'in Engineering', that a substantial percentage of courses should be taken within the student's technical competence.
3. *Professional electives* – supplementary technical managerial topics (about 30 per cent, 360 hours). Often the candidate finds that changing his job

responsibilities dictates needs and interests not in keeping with his previous background. Managerial and economic subjects will attract many.

4. *Outside interest electives* – humanities, cultural, etc. (about 20 per cent, 240 hours). We intend to encourage exploration of topics outside the traditional engineering experience in order to stimulate interest in a wider perspective, beyond the engineer's often narrow outlook.

It is important to further highlight three specially significant values that the programme includes for a candidate.

First, it is anticipated that most of the hours of credit will be earned through University of Wisconsin studies and its programmes, but to capitalize on many valuable opportunities at other universities, it is permitted that up to 600 hours (50 per cent of the total required) may be earned and transferred from courses and studies undertaken elsewhere. They must be well documented. Hours earned must be within the seven-year maximum allowed by the programme.

Second, it is required that each candidate, at some time during his programme engage in an individual Guided Special Study Project. About 200 hours of credit are recommended to be so earned. It should be a project of some worth, but the choice is wide. An on-the-job project may be ideal. Special faculty guidance will be assigned and a final report which is submitted for review is required as evidence of independent capability. The project will generally exist in the 'technical advancement' or 'professional elective' area.

Third, evidence of updating exists if the candidate becomes or is registered by examination as a 'professional engineer' – within the seven-year limit. For this, 140 hours of credit may be applied to the programme, but hours from 'Refresher' courses that may lead to registration become a part of the 140 total.

## PROCESS FOR THE INDIVIDUAL

When a student applies for acceptance to the programme he is asked to furnish information on his educational background and professional experience and to identify his objectives. An adviser from the faculty is assigned. He will assist in outlining an over-all programme to meet the objectives in line with economy and opportunities. In many cases he must help to define realistic objectives.

Taking background, experience, and the particular objectives into account, a programme spanning several years can be initially established. After review and agreement on the programme plans, the student is notified of formal acceptance as a candidate. He proceeds, then, to undertake the courses on the time-table established.

However, the student is not bound rigorously to all the details of the original plan. Considerable flexibility in the pursuit of an over-all programme is permitted and indeed is often desirable. New course opportunities can arise that were not initially available and may replace earlier choices. The student's job responsibilities and needs may change rather suddenly to alter some of his objectives. Between adviser and student, the broad programme can be adapted to these changes and opportunities.

Records of candidate progress are maintained by UW-Extension. P.D. degree programme policy and approvals are jointly administered by a committee of faculty representing both the College of Engineering at Madison and the Department of Engineering in Extension. When a student has completed the necessary minimum of 1,200 hours of credit (at least 1,000 on an evaluated basis), the faculty adviser will recommend to the P.D. degree committee that the candidate be approved for the degree. Upon this approval of the candidate's programme record the student may be awarded the University of Wisconsin Professional Development Degree in Engineering.

## EXPERIENCE OF THE PROGRAMME

Since the P.D. degree programme was announced in mid-summer 1970, it has attracted wide attention both in the industrial and public sectors and in the academic world. Over 1,300 inquiries have been handled, coming from industry, other universities, government and military agencies and hundreds of individuals. About 200 formal applications have been received. Some unqualified applicants have been rejected, others failed to pursue their initial interest.

Industry has shown that it is more than willing to allow a candidate time off for programme courses when made aware of the need sufficiently in advance. Industry's financial commitment for support of individual programmes has also been exceptionally favourable. Today there are almost 140 candidates actively under way on their P.D. degree programmes. A few have already completed more than half of the required total.

## CONCLUSION

Personal continuing professional education can run the gamut from regular reading of trade and professional periodicals to becoming involved in structured classroom studies. The time that should be invested will vary with each job discipline and with each individual, perhaps five hours per week to twelve hours per week. Few working engineers can maintain competence with less. Few can afford more.

The central idea embodied in the professional development degree concept is to make this continuing process efficient and effective for the engineer; to offer an educational programme that is comprehensive, disciplined and well-structured, yet flexible; to provide a programme that is one of quality; and to make attainable an achievement goal that will be recognized and respected.

# Diplôme de perfectionnement dans le domaine de la formation des ingénieurs (résumé)

Dr. John P. Klus

Ce document expose le programme du *Professional Development Degree* créé à l'Université du Wisconsin. Pour obtenir ce grade, le candidat doit totaliser 1 200 heures d'études homologuées ; il est prévu que, dans la plupart des cas, ces études dureront environ quatre ans. Le candidat aura besoin, par exemple, de suivre jusqu'à 240 heures de cours pour mettre à jour des connaissances oubliées ; il devra suivre au moins 360 heures de cours dans un domaine lui permettant d'augmenter sa compétence technique ; il devra vraisemblablement suivre environ 360 heures de cours dans des matières à option comme le droit ou la gestion des entreprises ; et il suivra peut-être environ 240 heures de cours dans des matières à option "extérieures" qui l'intéressent, comme les sciences sociales, les langues ou les beaux-arts.

Le concours de l'industrie du Wisconsin est attendu ; ce programme est considéré comme une nouvelle étape du développement de l'éducation continue et comme un effort qui mérite d'être connu et poursuivi.

# Survey of continuing education of engineers in FEANI countries

Osmo Ranta

## INTRODUCTION

The purpose of this rather superficial survey is to form a general picture of what is happening at the moment in this field in FEANI countries, to form an estimate of the volume of activity, and to collect some information on the problems, future plans, etc. 'as seen by the engineers' organizations.

For this purpose, questionnaires covering the national over-all picture were sent to the FEANI national member bodies, together with a set of questionnaires intended for individual institutions engaged in continuing education of engineers. The latter have been distributed by the national member bodies to representative institutions of their own selection.

The inquiry was limited to cover continuing education of engineers engaged in professional practice, excluding post-graduate education for the attainment of higher academic degrees, as well as pure management education for top managers.

The questionnaires were sent in February 1972 with a deadline set on 30 April, and some late replies were received until 10 August 1972 from fifteen national member bodies and from forty-five different institutions in fourteen countries.

The inquiry directed to national member bodies was divided into two parts, part A comprising a quantitative survey, and part B a qualitative survey describing the present situation and future plans in each country.

The questionnaires sent to different institutions included questions on the nature of organization, forms of activity, participants, volume of activity, educational means, policies relating to text material, and selection methods of course topics.

The results of the survey are summarized below.

## QUANTITATIVE SURVEY ON THE PRESENT SITUATION INDIFFERENT COUNTRIES

The quantitative part of the survey directed to national member bodies called for estimates of certain data which in many instances are not readily available from official statistics or elsewhere. Nevertheless these figures were thought to be of interest, so the national members were asked to fill in the respective figures even as rough guesses where more accurate information was missing. The results were surprisingly good, in this respect, so that only in a very few instances was a question left unanswered.

*Number of engineers*

The first question asked for an estimate of the number of engineers: (a) of university level (group A); and (b) of college or institute level (group B). In some cases only one group was reported, and this has been interpreted as corresponding to the total number of engineers in the respective countries. For the sake of brevity, the countries are indicated in the tables by letter combinations, which are identical to international motor-vehicle markings.

*Number of different institutions or organizational units offering continuing education for engineers in practice*

In this question, the following classification was used: 1. University or college extensions; 2. Institutions sponsored by engineer's organizations; 3. Institutions sponsored directly by the industry; 4. Government institutions; 5. Others. The results are shown Table 1.

*Table 1.* Number of different kinds of institutions offering continuing education for engineers in practice, by country

| Code | Country | Number of institutions in different categories | | | | |
|---|---|---|---|---|---|---|
| | | 1 | 2 | 3 | 4 | 5 |
| A | Austria | 2 | 1 | 3 | 1 | - |
| B | Belgium | 8 | 2 | - | - | 9[1] |
| ČS | Czechoslovakia | 14 | - | - | - | - |
| DK | Denmark | 1 | 2 | - | - | - |
| SF | Finland | - | 1 | - | - | - |
| D | Federal Republic of Germany | ... | ...2 | ... | ... | ... |
| GR | Greece | 2 | 1 | - | 1 | 4 |
| EIR | Ireland | 6 | 1 | 1 | 1 | - |
| L | Luxemburg | -3 | - | - | - | 1 |
| NL | Netherlands | 26 | 11 | 5 | 3 | 13 |
| N | Norway | 14 | 2 | - | 1 | 1 |
| E | Spain | 15 | 1 | - | - | 5 |
| S | Sweden | 3 | 4 | 2 | 1 | - |
| CH | Switzerland | 9 | 15 | 12 | 1 | 9 |
| GB | United Kingdom | ... | 8 | few | few | - |

1. Associations of educational institutions.
2. Data not given. Information available in reference 2 (see list at end of paper).
3. In preparation.
4. Not known, large number.

In the fifteen countries for which data are available, the numbers add up in different categories as follows:

| *Type of institution* | *Total number reported* |
| --- | --- |
| 1. University extensions | 100 (excl. GB) |
| 2. Institutions sponsored by engineers' organizations | 49 |
| 3. Institutions sponsored by industry | 24 |
| 4. Government institutions | 9 |
| 5. Others | 41 |
| Total number of institutions reported in fifteen countries | 183 |

*Approximate volume of activity*

In this part of the questionnaire, data were asked about the number of persons taking part in continuing-education activities (organized study) yearly, number of man-days spent in studies, and number of 'short' (less than sixty hours) and 'long' (more than sixty hours) courses given per year. The results are summarized in the following tables.

In this connexion it was also asked how the courses are divided into those intended for: (a) university graduates; (b) other engineers; and (c) mixed audience. The answers vary widely in this respect. In some cases all courses are reported as being intended for a mixed audience, in some cases the courses seem to be rather segregated for each category of engineers, the number of courses for mixed audience being small in proportion.

*Table 2.* Number of persons taking part in organized continuing-education activities yearly, by country. (Column 1. University graduates; Column 2. Other engineers; Column 3. Total; Column 4. Number of university graduates expressed as percentage of the total number of this category of engineers in each country)

| Country | Column 1 | Column 2 | Column 3 | Column 4 (%) |
| --- | --- | --- | --- | --- |
| A | 4,000 | 5,000 | 9,000 | 22 |
| B | 3,650 | 5,500 | - | 18 |
| ČS | 3,000 | - | - | 3 |
| DK | 2,500 | 6,500 | 10,200 | 25 |
| SF | 2,500 | 2,800 | 7,800 | 25 |
| D | - | - | 135,000 | - |
| GR | 1,600 | - | - | 10 |
| EIR | 500 | 50 | 550 | 17 |
| NL | 5,500 | - | - | 8 |
| N | 4,000 | 3,200 | 7,200 | 33 |
| E | 1,200 | - | - | 7 |
| CH | 6,000 | 6,500 | 12,500 | 30 |

*Note* : Column 3 (total number) is in some cases merely the sum of the figures in columns 1 and 2, in some cases larger, because it apparently includes non-engineers taking part in the same activities. Column 4, of course, gives a rough indication of frequency of attendance, but does not tell the time spent in studies.

*Table 3.* Volume of activity, by country. (Column 1. Approximate number of man-days spent in continuing engineering studies per year (only courses and other organized study); Column 2. Number of short or intensive courses (less than sixty hours) per year; Column 3. Number of long-duration courses (more than sixty lecture hours) per year)

| Country | Column 1 | Column 2 | Column 3 |
|---|---|---|---|
| A | 13,500 | 60 | 40 |
| B | - | 120 | 25 |
| ČS | - | - | 80 |
| DK | 27,000 | 563 | 12 |
| SF | 14,400 | 140 | - |
| D | ... [1] | 1,194 | 1,442 |
| GR | 10,000 | 17 | 8 |
| NL | 55,150[2] | 185 | 110 |
| N | 24,000 | 160 | 2 |
| E | 12,000 | 80 | - |
| CH | 40,000 | 350 | 100 |

1. 4.8 days per participant per year. Figures refer to courses within DVT (see reference 2).
2. University graduates only.

No criteria on the selection of events to be regarded as 'courses' were given in the questionnaire. Many replies, however, show that symposia, seminars, etc., have been left outside the figures. Possible criteria, as suggested in one reply (Netherlands), would be: (a) controlled attendance; (b) specified objectives; and (c) testing of educational performance. Of these (a) and (b) probably apply to most events reported but (c) would perhaps be too strict.

### *Sources of financing*

In this question, the national members were asked to indicate their estimates in rough percentages of total costs of continuing engineering studies coming from: 1. Employers; 2. Government funds; 3. Individuals; 4. Societies; 5. Others. The results are indicated in Table 4.

Estimates of this sort are, quite naturally, in most cases difficult to form. However, the differences ought to show clearly that different approaches to the financing problem exist. The employers, naturally, nearly always participate in financing by giving time off.

### *Influence on the selection of course topics*

In this part, the following alternatives were given: 1. National governing bodies (if any); 2. Industry; 3. Universities or institutions themselves; 4. Engineers in practice; 5. Engineering organization; 6. Professional societies; 7. Others.

The strength of influence was to be estimated according to a five-point scale: 1. Very weak; 2. Weak; 3. Average; 4. Strong; 5. Very strong.

The results are presented in Table 5.

*Table 4.* Sources of financing of continuing engineering studies, by country. Estimated rough percentage in money

| Country | Sources of financing (%) | | | | |
|---|---|---|---|---|---|
| | 1 | 2 | 3 | 4 | 5 |
| A | 20 | 20 | 40 | - | - |
| B | ...[1] | | | | |
| ČS | 50 | 50 | - | - | - |
| DK | 95 | 2 | 3 | - | - |
| SF | 99 | - | 1 | - | - |
| GR | - | 40 | 10 | 40 | 10 |
| EIR | 40 | 40 | 20 | - | - |
| NL | 37 | 15 | 36 | 12 | - |
| N | 90 | 5 | 5 | - | - |
| E | 50 | 25 | 25 | - | - |
| S | 95 | - | 5 | - | - |
| GB | ...[2] | | | | |

1. No data, all occur.
2. No data, large part directly or indirectly from government funds.

*Table 5.* Influence of various bodies on the selection of course topics (strength of influence given on a five-point scale)

| Country | Influencing body | | | | | | |
|---|---|---|---|---|---|---|---|
| | 1 | 2 | 3 | 4 | 5 | 6 | 7 |
| A | 2 | 4 | 3 | 3 | 5 | 4 | - |
| B | 1 | 4 | 3 | 3 | 4 | 4 | - |
| ČS | 3 | 4 | 3 | 2 | 1 | 1 | - |
| DK | 1 | 3 | 4 | 4 | 4 | 4 | - |
| SF | 1 | 3 | 2 | 4 | 3 | 3 | - |
| D | 1 | 5 | 3 | 5 | 4 | 4 | - |
| GR | 4 | 2 | 3 | 3 | 4 | 3 | - |
| EIR | 3 | 4 | 5 | 5 | 5 | 5 | - |
| NL | 1 | 4 | 3 | 5 | 4 | 2 | 1 |
| N | 2 | 4 | 2 | 3 | 3 | 3 | - |
| E | 3 | 2 | 4 | 3 | 3 | 1 | - |
| S | 1 | 4 | 2 | 3 | 4 | 3 | - |
| CH | 1 | 3 | 4 | 2 | 3 | 4 | - |
| GB | - | 4 | 5 | 3 | 4 | 2 | - |
| Total points | 24 | 50 | 46 | 48 | 51 | 43 | 1 |

Summing up the columns gives numbers of approximately the same size, except for columns 1 and 7. This shows that all selection mechanisms come into play in the countries considered except (7), i.e. the alternatives given were comprehensive. The influence of national governing bodies is, in general, considerably smaller than that of the other factors.

## QUALITATIVE SURVEY OF THE PRESENT SITUATION AND FUTURE PLANS

The national member bodies were asked to write short statements on the following points:

The need for continuing education for both groups of engineers.

The aim of continuing education for each group.

Timing of continuing education: (a) during working day or evening; (b) at what time during the engineer's career; (c) during unemployment.

Responsibility for the organizing of continuing education within: (a) the State; (b) educational institutions; (c) industry; (d) engineering organizations; (e) others.

Special legislation covering continuing professional education.

Future plans: (a) filling the gaps in continuing education; (b) improvements in prospect; (c) improvements which could be suggested.

Many member bodies have sent in very clear and lucid statements which sum up the situation in their respective countries at the moment, as seen by the engineers' organizations. These statements are briefly summarized below.

In the case of the Federal Republic of Germany, two recent publications exist describing the problems of continuing education in science and engineering [1][1] and giving detailed statistics on the lively activity of the scientific and engineering societies of the Federal Republic in this field [2]. Similarly, Svenska Teknologföreningen in Sweden has recently published a book presenting problems of continuing engineering education and suggesting solutions [3].

For France and Greece, reference is made to the papers of Professors Ferry and Tassions (Annex 2).

Some salient points in the replies are presented in the following short summary. The origin of certain statements is indicated by a letter code (see Table 1) referring to countries. From some countries several replies were received, presenting the views of different organizations.

### *Need for continuing education*

The need seems to be recognized almost universally, and no general need for increased publicity effort seems to exist.

(B) Nobody has to be convinced of the necessity of continuing education.

(DK) Closer contact with the market is desirable in order to cater for the qualitative needs. Effort is made to offer advanced courses in new subjects. A tendency among the companies to save on education costs is observable, due to weakening of the economy as compared with the sixties.

1. The figures in brackets refer to the references at the end of this article, page 70.

(NL) The basic education of engineers should be restricted to fundamentals. The need for continuing education will grow from this reason, and also because of the explosive development of science and technology.

(N) Continuing education is an integrated part of the engineer's total education and is a lifelong process. In 1969, the estimated time used for continuing education in Norway was about 1 per cent of total working hours. Five per cent should be used in the near future (estimate of a government committee).

(GB) The number of engineering graduates being produced from United Kingdom universities and polytechnical colleges at the moment is such that, within the next ten years, career structures within industry will change considerably. The group A class of engineer may have to seek a career normally reserved for the group B engineer. The need for continuing education is equal in both groups.

*The aim of continuing education*

The engineer's own specialization was generally reported to be the main sphere of interest in continuing education. Other technical disciplines come into play as interdisciplinary and borderline interests. Non-technical disciplines such as management and economics are quite generally reported to be of great interest to engineers. A pressing need is felt (GB) for a type of continuing education that is suited to the future needs of industry. The Open University (GB) and Teleac (Television Academy, NL) offer good prospects for study in non-technical disciplines.

*Timing of continuing education*

In nearly all replies the main emphasis was put on education taking place during the working day, but evening and weekend study are not to be excluded. An interesting prospect is a four-day week, which could become a practical reality in advanced industrial societies; it would then be possible to use the fifth day for continuing education and maintain the leisure sanctity of the weekend.

As to the time during the engineer's career when continuing education should mostly be given, the replies vary. Many replies state that all ages are possible, whereas others place the main emphasis on engineers having up to fifteen years of practice.

Unemployment is generally not considered a significant factor, but a potential need is felt here if the employment situation should grow worse. In that case, re-education would become a practical necessity.

*Special legislation*

Special legislation covering continuing education was reported from Czechoslovakia. Another example is France.

*Responsibility for the organizing of continuing education*

The replies indicate that the engineers' organizations bear a heavy responsibility, notably in the Scandinavian countries (DK, SF, N, S). In some countries educational institutions take the main load but the general tendency is in all countries for this part to increase. The State takes its responsability mainly indirectly through the funding of educational institutions.

*Future plans*

(A) Public recognition of the needs of continuing education is still too weak and the situation must be improved.

(B) No plans for a more advanced organization of continuing education exist. The educational institutions have other pressing problems, and the problems of continuing education are thus set aside.

(ČS) Some organizational problems exist, and these are being investigated at the moment. The aim is to set up a specialist's degree on top of the basic university degree and to organize a short-course activity.

(DK) The Danish engineers' organizations (Dansk Ingeniørforening (DIF) and Ingeniør-Sammenslutningen (IS)) have given separate replies. DIF hopes to improve the contact with management in all sectors of society. Co-operation with educational institutions is to be widened, and international co-operation with similar organizations is to be strengthened. Teachers are to be attracted increasingly from other countries. It also hopes to widen international co-operation, and tries to improve the educational skills of teachers in continuing-education institutions in different countries.

(EIR) Introduction of polytechnic type courses.

(NL) Further education, updating education and re-education should be arranged systematically in the future through voluntary co-operation between government, commerce, industry, educational institutions and engineers' organizations.

(N) Co-operation between industry, universities, technical colleges and engineers' organizations is essential for a well-planned continuing education. Committees are suggested to plan the continuing education at the various educational centres (universities and technical colleges).

(GB) The Open University is a great advance in continuing education, which would seem to be capable of great developments in this field.

## INFORMATION ON THE ACTIVITIES OF CERTAIN REPRESENTATIVE INSTITUTIONS

The FEANI national member bodies were asked to forward a questionnaire covering details of the activities to representative institutions engaged in continuing education of professional engineers in their countries. No specific directions were given as to the selection of institutions, and no data are on hand on the total number of institutions which received questionnaires. The number of replies was forty-five and the institutions concerned are divided between fourteen countries. It must be emphasized that the results of this part of the survey merely list some examples, and that especially big countries with a large total volume of activity are rather under-represented.

*List of institutions replying to the survey*

A list of the names and addresses of the institutions is given in Table 6. In the subsequent text, the institutions are referred to by a code indicating country and institutions.

Among the institutions from which replies were received, are 22 university or college extensions; 14 institutions sponsored by engineers' organizations;

*Table 6.* Some institutions engaged in the continuing education of professional engineers in FEANI countries.

| Code designation | Name and address of institution | Nature of institution[1] | Forms of activity[2] | Participants[3] |
|---|---|---|---|---|
| B1 | Faculté Polytechnique de Mons, Rue de Houdain 9, B-7 000 Mons | 1 | 1, 2, 3 | 1 |
| B2 | Katholieke Universiteit Leuven, Faculteit toegepaste wetenschappen (Faculty of Applied Sciences), 94 Kardinaal Mercierlaan, B-3 030 Heverle | 1 | 1, 2 | 1, 2 |
| B3 | Université Catholique de Louvain, Faculté des Sciences Appliquées, Château d'Arenberg, B - 3 030 Heverle | 1 | 3 | 1, 2 |
| B4 | Koninklijke Vlaamse Ingenieursvereiniging (Royal Flemish Engineering Society), 58 Jan van Rijswijcklaan, B-2 000 Antwerpen | 2 | 1, 3 | 1, 2, 3 |
| B5 | Faculté des Sciences Agronomiques, B-5 800 Gembloux | 1 | 2, 5 | 1 |
| B6 | Université Libre de Bruxelles, Faculté des Sciences Appliquées, 50 Avenue F.-D. Roosevelt, B-1 050 Bruxelles | 1 | 2, 3 | 1 |
| B7 | Ecole Royale Militaire, Avenue de la Renaissance 30, B-1 040 Bruxelles | 5 | 2 | 1 |
| DK1 | Dansk Ingeniørforening–Efteruddannelsen, Langebjerg 4, DK-2 850 Naerum | 2 | 1, 2, 3 | 1, 2, 3 |
| DK2 | Ingeniør-Sammenslutningen, Efteruddannelsen, Ved Stranden 18, DK-1 061 Copenhagen K | 2 | 1, 2, 3, 4 | 1, 2, 3 |
| SF1 | Insinöörijärjestöjen Koulutuskeskus – Ingenjörsorganisationernas Skolningscentral, Annankatu 18, SF-00120 Helsinki 12 | 2 | 1 | 1, 2, 3 |
| F1 | Ecole Supérieure d'Electricité, 10 Avenue Pierre Larousse, 92 Malakoff | 2, 3 | 1, 2, 3 | 1, 2, 3 |
| F2 | Ecole Nationale Supérieure des Mines de Paris, 60 Boulevard Saint Michel, Paris-6ᵉ | 4 | 1, 2, 3 | 1, 2, 3 |
| F3 | Centre de Formation Continue ENSEEIHT, Ecole Nationale Supérieure d'Electrotechnique, d'Informatique et d'Hydraulique de Toulouse, 2 Rue Charles Camicher, 31 Toulouse | 2 | 1, 2, 3, 4 | 1, 2, 3 |
| F4 | Ecole Nationale Supérieure de l'Aeronautique et de l'Espace | 4 | 1, 2 | 1, 2 |

| | | | | |
|---|---|---|---|---|
| D1 | Fachhochschule Furtwangen, Gerwigstrasse 11, 7743 Furtwangen (Schwarzwald) | 1 | 1, 2, 3 | 1, 2, 3 |
| D2 | Ruhr-Universität Bochum, Institut für konstruktiven Ingenieurbau, Fachbereich Informationsverarbeitung, Buscheystr. 1 64/147 | 1 | 1, 2, 3 | 1 |
| D3 | Institut für Kernenergetik der Universität Stuttgart, Pfaffenwaldring 31, 7 000 Stuttgart 80 | 1 | 2 | 1 |
| D4 | Technische Akademie e.V., D-56 Wuppertal 1 | 1 | 1, 2, 3 | 1, 2, 3 |
| EIR1 | Trinity College, Dublin | 1 | 1, 2, 3 | 1, 2, 3 |
| EIR2 | Mechanical Engineering Department, University College, Upper Merrion Street, Dublin | 1 | 2 | 1 |
| I1 | Facoltà d'Ingegneria del Politecnico di Torino, Corso Duca degli Abruzzi 24, 10 129 Torino | 1 | 1, 2, 3 | 1, 2, 3 |
| NL1 | Training Division of SIPM (Shell), P.O. Box 162, The Hague | 5 | 1, 2, 3, 4, 5 | 1 |
| NL2 | 23 colleges of technology (joint answer), c/o Nieuwe Parklaan 11, The Hague | 1 | 2 | 1 |
| NL3 | Opleidingscentrum PTT 'Voorlinden' Groep Loopbaancursussen, Buurtweg 90, Wassenaar | 4 | 1, 2, 3 4, 5 | 1 |
| NL4 | Stichting Post-Doktoraal Onderwijs in het Bouwen, Technische Hoogeschool Delft, Berlageweg 1, kab', 303, Delft | 1, 2, 5 | 1, 3 | 1 |
| N1 | The Norwegian Society of Professional Engineers, Continuing Education Division, Kronprinsensgate 17, N-Oslo 2 | 2 | 1, 3, 4 | 1, 2, 3 |
| N2 | Norges Ingeniørorganisasjon – NITO, Kursavdelingen, P.O. Box 7 062, Homansbyen, N-Oslo 3 | 2 | 1, 2, 4 | 1, 2 |
| P1 | CEGOC – Centro de Estudos de Gestão e Organizacao Cientifica, Rua Castilho 167, 1° Lisboa | 2 | 1 | 1, 2, 3 |
| GB1 | Paisley College of Technology, High Street, Paisley, Renfrewshire, Scotland | 1 | 1 | 1, 2, 3 |
| GB2 | Sheffield Polytechnic, School of Engineering (3 departments), Pond Street, Sheffield, S1 1WB | 1 | 1, 2, 3 | 1, 2, 3 |

| Code designation | Name and address of institution | Nature of institution[1] | Forms of activity[2] | Participants[3] |
|---|---|---|---|---|
| GB3 | Manchester Polytechnic, Chester Street, Manchester | 1 | 1, 2 | 1, 2, 3 |
| GB4 | Bristol Polytechnic, Department of Engineering, Ashley Down Road, Bristol, BS7 9BU | 1 | 1, 2, 3 | 1, 2 |
| GB5 | Robert Gordon's Institute of Technology, School Hill, Aberdeen, AB9 1FR, Scotland | 1 | 1, 2, 3 | 1, 2 |
| GB6 | The Institution of Mechanical Engineers, 1 Birdcage Walk, London, S.W.1 | 2 | 1 | 1 |
| S1 | Svenska Teknologföreningen STF, Ingenjörsutbildning, Malmskillangsgatan 48 A$^{IV}$, S-103 43 Stockholm 40 | 2 | 1 | 1, 2, 3 |
| S2 | Swedish Employers, Confederation, Educational and Training Department, ALI-RATI Industrial Management Training Institute | 3 | 1, 2, 5 | 1, 2, 3 |
| S3 | Svenska Väg-och Vattenbyggares Riksförbund (Swedish Society of Civil Engineers), Regeringsgatan 98, S-111 39 Stockholm | 2 | 1 | 1 |
| S4 | Svenska Byggnadsingenjörenas Riksförbund SBR, Teatergatan 3, S-111 48 Stockholm | 2 | 1 | 2 |
| S5 | Chalmers University of Technology, Kurssekretariat CTH, Fack, S-402 20 Göteborg 5 | 1 | 1 | 1, 2 |
| CH1 | Ecole Polytechnique Fédérale de Lausanne, 33 Avenue de Cour, 1 007 Lausanne | 1 | 2, 3 | 1 |
| CH2 | Seminar für technische Weiterbildung CH-8 400 Winterthur | 4 | 1, 3 | 1, 2, 3 |
| CH3 | Höhere Technische Lehranstalt, Brugg-Windisch, CH-5 200 Windisch | 1 | 1, 2, 3 | 1, 2 |
| CH4 | Eidgenössische Technische Hochschule, Laonharstrasse 33, 8 006 Zürich | 1 | 1, 2, 3 | 1, 2, 3 |

1. 1 = University or college; 2 = Institution sponsored by engineers' organization; 3 = Institution sponsored by industry; 4 = Separate government institution; 5 = Other.

2. 1 = Short courses (less than sixty lecture hours); 2 = Long courses (more than sixty lecture hours); 3 = Periodic instruction (e.g. lecture series); 4 = Organized self-study (correspondence courses); 5 = Other.

3. 1 = Engineers of category A of the FEANI register (university level); 2 = Engineers of category B of the FEANI register (institution level); 3 = Other professional groups.

4 institutions sponsored by industry; 3 separate government institutions; 4 institutions of another nature.

Some institutions fall into two or three of these categories.

*Forms of activity*

The reported forms of activity (fourth column) are distributed among the replies as follows: short courses in 37 cases; long courses in 33 cases; periodic instruction in 25 cases; organized self-study in 6 cases; other in 5 cases.

Under the last such forms of activity as on-the-job training (NL1) and individual programmes (NL3) were reported.

*Participants*

Of the 43 institutions listed, 14 offer instruction to university engineers only; 1 offers instruction to group B engineers only; 8 offer instruction to both group A and B engineers; 20 offer instruction to engineers and other professional groups.

In the last case, groups having had less formal schooling than engineers were reported (engineering technicians, foremen, etc.), as well as groups having the same level of basic education (medical doctors, geologists, etc.).

*Volume of activity*

The reported volume of activity varies greatly. Limits of variation in the replies showed that 8 to 18,000 persons tcok part in 2 to 800 courses per year.

Detailed statistics for the volume of activity were given in most replies, whereas in many cases, quite understandably, such data were not readily available. The institutions reporting short courses, or short and long courses, fall into groups according to the number of separate courses given per year as follows:

between 1 and 10 : B2, D1, EIR2, NL4, GB1, GB3, GB5, S5, CH2
between 11 and 100 : F1, F2, F3, I1, NL1, NL2, NL3, N2, P1, GB2, GB4, GB6, S3, CH4
101 or more: B4, DK1, DK2, SF1, D4, N1, S1, S2

The replies indicate rather clearly the obvious fact that the institutions offering long specialist courses lasting in some cases a year or even more limit themselves to a very small number of courses and also to a small number of participants.

*Educational methods and means used*

The institutions concerned were also asked about educational means and methods used. The following categories were included in this question: 1. Lectures; 2. Exercises (problem-solving, etc.); 3. Laboratory work; 4. Printed textbooks; 5. Other texts; 6. Self-instructing material (programmed books, etc); 7. Television instruction (closed-circuit television).

The replies are summarized in Table 7.

Among other methods and means used, film was reported in several instances. Other methods reported were case studies (GB2), group dynamic methods (S2), small tutorials (GB4), and factory visits (F1).

In the case of self-instructing materials, slide-tape programmes were mentioned in one case (NL1).

*Table 7.* Educational methods and means used

| Method | Number of replies | | | | |
|---|---|---|---|---|---|
| | Not at all | Occasionally | Usually | Extensively | Exclusively |
| 1. Lectures | - | 3 | 15 | 25 | - |
| 2. Exercises | - | 6 | 17 | 18 | - |
| 3. Laboratory work | 5 | 14 | 12 | 7 | 1[1] |
| 4. Printed textbooks | 8 | 11 | 12 | 5 | 1 |
| 5. Other texts | - | 8 | 14 | 14 | 3 |
| 6. Self-instructing material | 16 | 16 | 3 | - | - |
| 7. Television instruction | 21 | 11 | - | - | - |

1. Individual work projects (Aufbaustudien) (D3).

*Teachers used*

The next question asked for information on the teaching staff, using the following categories: 1. Research workers; 2. Professional teachers (university teachers,etc.); 3. Practising engineers; 4. Others.

The results are presented in Table 8.

*Table 8.*

| Teacher category | Number of replies | | | | |
|---|---|---|---|---|---|
| | Not at all | Occasionally | Usually | Extensively | Exclusively |
| 1. Research workers | 1 | 24 | 8 | 5 | - |
| 2. Professional teachers | - | 8 | 12 | 18 | 4 |
| 3. Practising engineers | 1 | 14 | 15 | 10 | - |
| 4. Others | 1 | 6 | 1 | 1 | - |

Among the last group, of course, fall various experts in other fields. Specific groups mentioned in the replies were accountants, marketing specialists, economists, lawyers, architects, managers, trade unionists.

A significant trend in the replies is perhaps the indication of roughly equal prominence of professional teachers and practising engineers. This reflects the fact that university extensions and similar institutions, that already have full-time teaching staff, use this for the education of engineers who are working in the field.

*Policies relating to text material*

Two questions were asked in this connexion concerning (a) distribution of material for advance study; and (b) material available for sale to non-participants.

The replies are indicated in Table 9.

*Table 9.* Policies relating to text material

| | Number of replies | | | | |
|---|---|---|---|---|---|
| | Not at all | Occasionally | Usually | Extensively | Exclusively |
| Material is distributed for advance study | 4 | 14 | 19 | 3 | 1 |
| Material is available for sale to non-participants | 17 | 15 | 4 | 4 | 1 |

### *Selection of course topics*

Mechanisms and methods of selection of topics to be taught were investigated by four questions, with results as shown in Table 10.

*Table 10.* Selection of course topics

| Selection of course topics | Number of replies | | | | |
|---|---|---|---|---|---|
| | Not at all | Occasionally | Usually | Extensively | Exclusively |
| Is influenced by individuals | 2 | 12 | 14 | 9 | 1 |
| Is influenced by expert groups | 3 | 12 | 10 | 12 | 3 |
| Is based on inquiries to industry | 4 | 16 | 9 | 10 | - |
| Is based on inquiries to participants | 5 | 10 | 15 | 8 | 1 |

## CONCLUSION

It was clear from the beginning that it was impossible to make a truly representative let alone complete survey of this vast activity in eighteen countries, involving well over 1.5 million engineers. Yet it was felt that the experience of earlier seminars indicates a considerable interest in international comparisons, and so the author hopes that this modest survey may serve as a starting-point for further discussions and exchanges of ideas.

Because the continuing education of engineers is a costly business and is in many cases limited by resources of different kinds, it would be important to shed more light on the factors affecting the efficiency of this activity. For that purpose methods of sociological and educational research should be used. In fact, such studies have been made, often in the form of graduate theses, but little has been published so far.

Thanks are due to the FEANI national member bodies for their benevolvent attitude and their valuable contributions to this survey.

The institutions sending in replies have also done much work in filling the questionnaires, which is gratefully acknowledged. Many institutions have sent samples of education programmes and other material.

*References*

1. 'Weiterbildung und Lebenslanges Lernen in den Naturwissenschaftlichen und Technischen Berufen. Ergebnisse der Jahresversammlung 1970 des Deutschen Verbandes Technischwissenschaftlichen Vereine (DVT)', *VDI Information,* no. 24, November 1970, 31 p. (In German.) This publication presents six articles on continuing education of scientists and engineers, besides discussion during the annual meeting of the German Federation of Scientific and Technical Societies (DVT), and recommendations made by the meeting.
2. 'Die Arbeit der Technisch-wissenschaftlichen Vereine auf den Gebiet der Weiterbildung. Eine Erhebung bei den Mitgliedgesellschaften des DVT', *VDI Information,* no. 26, March 1971, 14 p. (In German.) Results of a survey made among the sixty-one member societies of DVT on continuing-education activities.
3. *Ingenjörerna och Fortbildingen.* Stockholm,Svenska Teknologföreningen, 1971, 56 p. (In Swedish.) Report of a committee set up by Svenska Teknologföreningen on the problems of continuing education of engineers. Includes suggested solutions.

# Etude sur la formation permanente des ingénieurs dans les pays de la FEANI (résumé)

Osmo Ranta

Ce document donne les résultats d'une étude effectuée sur la formation permanente des ingénieurs dans les pays de la FEANI. Après une estimation du nombre des ingénieurs dans chaque pays, des données sont fournies sur le nombre des institutions s'occupant de formation permanente des ingénieurs, ainsi que sur celui de personnes prenant part aux activités organisées dans ce domaine. Viennent ensuite des renseignements complémentaires sur la durée des cours, leurs sujet et leur financement, puis une discussion générale sur le besoin de formation permanente, les buts et l'organisation de celle-ci.

# General discussion

Among the questions raised in the time available for discussion was one concerning the role of the colleges of advanced technology in Australia. The speaker asked whether they were making a contribution to the continuing education of engineers comparable to their contribution to the initial education of engineers. In reply, it was stated that they were not involved in the continuing education of engineers to any great extent.

The summer institutes for engineering teachers that were initiated in India in about 1964 as an international co-operative project were mentioned. These institutes have subsequently flourished, and the state and federal governments are now participating. It was stated that a number of different forms of this scheme are now used, including sequential summer schools.

A speaker confirmed that in the U.S.S.R. the Ministry of Higher and Specialized Secondary Education has the entire responsibility for creating any special institution for the continuing education of engineers. He also mentioned that courses are normally of at least two months' duration, rather than the much shorter duration often preferred in other countries, and that participants are sent by their employers. He said that university staff are often sent on such courses.

The same speaker went on to describe the situation in Poland, and mentioned that it was proposed there that continuing education be compulsory. This provoked considerable discussion, one speaker mentioning that the motivation was an important factor and that even in Japan it appeared that only 6 per cent of the engineers participated in continuing-education courses. The question was also raised as to whether the engineer alone should decide on whether he should receive continuing education or whether society as a whole should make this decision. Recent Californian legislation requiring professionals to keep their skills current was mentioned.

One speaker suggested that engineers may in some instances be exploited by their employers, and in fact were not being given the opportunity to keep up to date.

The work in depth of OECD in this area was described. Another speaker mentioned the situation in Venezuela, where continuing education has been controlled by the professional societies and a charter foundation has been established to be responsible for its support.

The situation in the United Kingdom was described, and the contributed paper on this subject summarized. Another speaker elaborated on this and described the 'total technology' scheme for career development of promising young engineers that was being promoted as a less-academic alternative to Ph.D. studies.

The figures given in the survey of FEANI countries were queried by several speakers and it was indicated that this initial survey was incomplete and to some extent superficial and should be followed by more careful studies.

Messrs Barry, Bose, Corbett, Goodyear, Ham, King, Klus, Leonard, Mizushina, Ranta, Tassios, Ter-Davtian, Titchener, Tymowski and Yackovlev contributed to the discussion.

# Main plenary papers presented on 22 August 1972

# Principaux documents présentés en séance plénière le 22 août 1972

# III

# La formation continue de l'ingénieur en France

Marc Pelegrin

## PRINCIPES

Dans le désarroi qui frappe la jeunesse actuelle, un élément reste sûr : c'est le désir qu'elle manifeste d'accroître ses connaissances, soit intrinsèquement pour mieux comprendre le monde dans lequel elle vit, soit professionnellement pour mieux s'adapter aux métiers ou pour s'élever dans la hiérarchie des fonctions.

Mais la complexité actuelle des techniques, au sens large du terme – complexité qui ira croissant – interdit de donner un bagage intellectuel de base définitif durant l'adolescence. De plus, malgré les efforts des gouvernements pour démocratiser l'enseignement en multipliant les bourses ou prêts d'honneur, aux étudiants les moins favorisés, il faut tenir compte des aléas de la vie qui peuvent contrarier les projets des jeunes et les obliger à travailler plus tôt qu'ils ne l'auraient souhaité.

La formation continue, dont nous étudierons les caractéristiques plus tard, permet de satisfaire le besoin de connaissance et d'acquérir ou de renouveler les aptitudes professionnelles, un peu au gré des penchants de l'industrie. On trouve donc, dans la formation continue, un facteur de justice et de progrès tant pour l'individu que pour la société.

Mais c'est également une condition désormais indispensable pour le développement économique d'un pays. Les méthodes actuelles de gestion d'un atelier, d'un magasin, d'une société d'assurances sont totalement différentes de celles d'il y a vingt ans et pourtant ce sont en général des hommes de quarante ans qui assument les responsabilités principales. Certes, leur expérience leur aura permis une certaine évolution mais à quel prix ? Il y a vingt ou trente ans la formation et l'entretien des pilotes de ligne se faisaient essentiellement en vol ; actuellement, près de la moitié de la formation (et de l'entretien) se passe sur simulateur électronique. Les nouveaux procédés de navigation ou de pilotage sont d'abord appris sur simulateur afin que la phase d'instruction en vol soit limitée au maximum. Qu'on mesure l'importance de l'effort de formation ainsi consenti en remarquant que le prix d'un simulateur est souvent supérieur à la moitié du prix d'un avion neuf ! C'est la compétition économique qui est la principale cause de cette évolution, mais il faut noter que le simulateur permet de faire plus que l'avion : ainsi, on prendra rarement le risque de couper un propulseur sur un bipropulseur durant la phase de décollage, alors que l'entraînement du pilote sur simulateur peut se faire dans les configurations les plus anormales. Il y aurait beaucoup à dire sur le rôle des simulateurs dans la formation technique future. Plusieurs compagnies envisagent actuellement de supprimer totalement les vols d'entraînement sur nouvel avion : les pilotes devant être transformés effectueront d'abord un grand nombre d'heures sur simulateur, puis passeront directement en copilotes sur lignes commerciales.

C'est, bien entendu, dans le domaine des techniques que cette évolution est la plus rapide. C'est dans ce domaine que le recyclage ou l'enseignement permanent doit être le plus actif.

Mais c'est aussi dans ce domaine que la variété et la diversité des enseignements à offrir sont les plus vastes. C'est également dans ce domaine que la souplesse et la rapidité de mise en place de nouveaux enseignements sont les plus nécessaires.

Dans la suite de cet exposé, nous considérons principalement le recyclage dans le domaine technique et plus particulièrement celui des ingénieurs (140 000 diplômés sont actuellement en fonction ; 9 000 environ sont formés par les écoles chaque année).

## LA FORMATION CONTINUE EN FRANCE

Au cours des dernières années, une série de dispositions prises par le gouvernement et les grandes organisations patronales et syndicales donnent à la France une place privilégiées dans la formation et le perfectionnement professionnels.

On peut situer, un peu arbitrairement, en mai 1968 le début de négociations serrées entre les syndicats et le gouvernement. Des accords interprofessionnels ont été signés en 1969, 1970 et 1971, et la loi du 16 juillet 1971 – qui reprend les dispositions principales de cet accord – fixe les règles de la formation professionnelle pour les jeunes et les adultes dans un contexte général.

On trouvera en annexe une analyse sommaire de ces textes relatifs à la formation continue, plus particulièrement en ce qui concerne la formation des ingénieurs et cadres supérieurs.

Pour l'ingénieur, l'accord du 30 avril 1971 et la loi du 16 juillet affirment la nécessité du recyclage et prévoient la mise en place de moyens accrus.

## UN EXEMPLE DE FORMATION CONTINUE A L'ENSAE

L'Ecole nationale supérieure de l'aéronautique et de l'espace (Sup'Aéro) a sans doute été parmi les premières, sinon la première école, à offrir dans le cycle scolaire (octobre à juin) une série de thèmes, pas uniquement centrés sur l'aéronautique d'ailleurs, pour le "recyclage" des ingénieurs. C'est en effet en 1958 que l'ingénieur général Lehr a pris en charge cette responsabilité dans le cadre de la Société des amis de l'école (SAE) car, comme nous l'avons dit plus haut, seule une structure offrant une grande souplesse et assurant la mise en place des décisions prises dans des délais très courts – au maximum une année – peut remplir cette mission.

Les structures administratives des grands services de l'Etat – dont dépend l'école – n'ont pas un temps de réaction aussi court et sont en général régis par des textes très stricts quant aux budgets, aux postes de professeurs, aux investissements des laboratoires.

Une structure purement privée et commerciale a pour objectif premier celui de la rentabilité des investissements réalisés, et ce souci peut estomper le but fondamental de cette forme d'enseignement, qui est d'accroître le niveau des connaissances sans que nécessairement en résulte un bénéfice économique direct pour la société.

Une structure privée, non commerciale, c'est-à-dire sans but lucratif ([1]) (association du type "loi 1901"), étroitement liée à une école, nous paraît être la structure optimale. Les statuts de l'école lui reconnaissent une mission "d'enseignement postscolaire", mais précisent que cette mission peut être placée sous la responsabilité d'un organisme extérieur moyennant l'établissement de "conventions" passées entre cet organisme et l'administration.

La définition des enseignements à dispenser est faite par la direction de l'école (principalement par le directeur des études) en fonction des demandes de l'industrie pour le court terme et en fonction des estimations de l'évolution des techniques pour l'avenir à plus long terme (fig. 1).

Le directeur des stages de la SAE établit le coût du stage dans l'hypothèse d'un nombre d'inscrits fixé d'un commun accord par l'école et la SAE. A la fin du cycle scolaire, on examine quel a été le degré de participation à chaque stage, ce qui constitue un élément important pour fixer la politique de l'année suivante. Mais il est intéressant de noter que souvent des stages déficitaires sont maintenus pendant un an, voire deux.

La direction de l'école peut estimer que, si le stage n'a pas recueilli le nombre escompté d'auditeurs, c'est que : *a*) sans que son contenu soit à mettre en cause, son titre, sa présentation ou son programme détaillé sont à revoir car "ils ne plaisent pas" ; *b*) les ingénieurs n'ont pas eu conscience de l'importance de cette technique, et seule la propagande directe qui sera faite par les quelques auditeurs qui ont suivi le stage permettra à celui-ci d'atteindre ou de dépasser son seuil de rentabilité un ou deux ans plus tard.

Qui enseigne et comment enseigne-t-on ? Comme à l'école, ce ne sont pratiquement que des ingénieurs qui enseignent aux futurs ingénieurs et nous estimons que cette façon d'enseigner – assez rare dans le monde – est à maintenir. Si, en effet, le contact est plus direct, car les origines sont les mêmes, il est, de plus, important que l'enseignement donné à des ingénieurs, dont la moyenne d'âge est actuellement de trente-cinq à quarante ans, ne soit pas un enseignement de type scolaire ou universitaire. L'ingénieur qui enseigne (à titre de fonction accessoire) sait de quoi il parle et, même s'il ne sait pas discourir avec tout le brio d'un professeur de faculté, il transmet mieux son savoir, ne serait-ce que par le langage qu'il utilise.

L'enseignement de recyclage pour les ingénieurs doit nécessairement être assorti de nombreux exercices d'application et, chaque fois que c'est possible, de travaux de laboratoire. Nous nous efforçons, à Sup'Aéro, de respecter la règle d'une heure d'exercices ou de travail de laboratoire pour chaque heure de cours. C'est pourquoi nous disions que la présence d'un centre de recyclage auprès d'une école est une excellente chose : c'est l'évidence même pour les laboratoires, mais c'est tout aussi vrai pour les exercices, car il s'agit souvent de problèmes d'une durée de trois heures, voire six heures ; leur préparation et leur mise au point demandent à l'auteur entre dix et trente heures suivant le sujet traité. Le professeur (ingénieur) donne en général un cours similaire à l'école et il peut, de ce

1. "Non-profit organisation" aux Etats-Unis d'Amérique

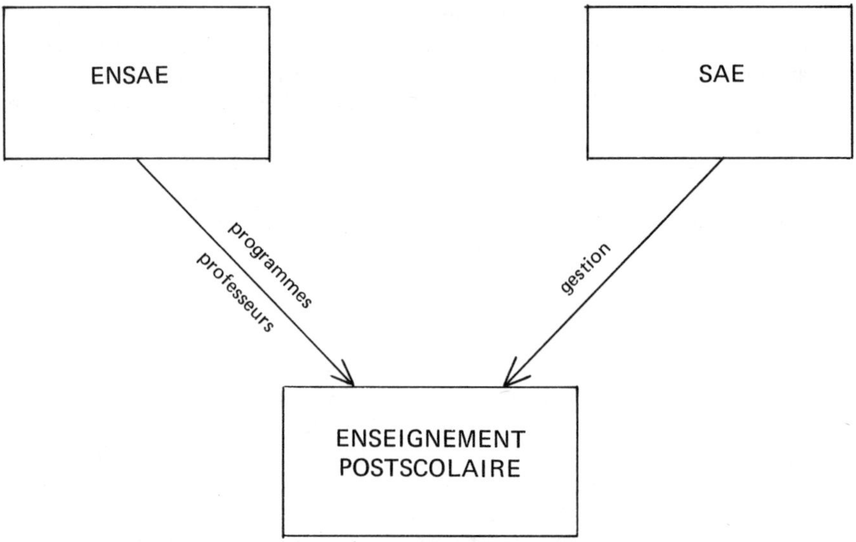


fait, bénéficier du lot de problèmes, exercices et travaux pratiques déjà mis au point pour les élèves.

Je mentionnerai, parallèlement à ces stages de recyclage, dont la durée varie de une à quatre semaines, des enseignements d'une année, donnés directement par l'école sous l'appellation de "années de spécialisation".

Rappelons d'abord que Sup'Aéro, comme ses consoeurs (Ecole des mines, Ecole des ponts et chaussées, Ecole des télécommunications, Ecole des techniques avancées) ont pour mission principale de former des ingénieurs généralistes destinés à concevoir les matériels ou ouvrages nouveaux. L'enseignement se veut aussi général que possible dans la discipline de l'école. (Notons que cette condition est pratiquement impossible à satisfaire dans la plupart des techniques : ainsi, en 3e année de Sup'Aéro, les élèves se répartissent en trois options : Avions et missiles, Propulsion, Avionique. Ils reçoivent tous – s'ils le méritent – le même diplôme d'ingénieur de l'aéronautique).

Afin d'accélérer la formation complémentaire que l'ingénieur acquiert normalement dans l'usine même où il travaille, on a créé, il y a six ans, quatre thèmes de spécialisation s'étendant sur une année (Mécanique aérospatiale, Informatique, Automatique avancée, Systèmes). Ces cours sont ouverts à tous les ingénieurs diplômés ; les admissions sont prononcées par un jury.

L'expérience montre que rares sont, pour le moment, les ingénieurs qui, ayant plusieurs années d'expérience, s'inscrivent à ces spécialisations (qui pourtant sont gratuites). Par contre, les jeunes ingénieurs inscrits (une cinquantaine) viennent d'horizons divers : la proportion d'élèves de Sup'Aéro venant d'être diplômés n'excède pas 20 %.

## PERSPECTIVES D'AVENIR ET CONCLUSION

La rapide croissance des effectifs bénéficiant de ces stages de recyclage depuis 1960 montre qu'en dehors de toute loi nouvelle telle que celle de juillet 1971, l'enseignement de juillet 1971 crée une situation nouvelle qui va se traduire par une brusque accélération de cet accroissement.

Lorsque le régime transitoire sera terminé, ce sera une masse monétaire de 5 à 10 milliards de francs qui sera en jeu. Les estimations les plus pessimistes évaluent à 20 % le volume réservé aux "disciplines de l'ingénieur". Or, dans la situation actuelle, on estime à 20 millions de francs les sommes mises en jeu en 1971/72 par le recyclage (environ 200 000 journées-stagiaires ont été données ; leur prix moyen semble être de 100 francs). Il s'agit donc, en France, de multiplier par 100 l'effort fait pour le recyclage et l'enseignement permanent des ingénieurs.

L'ampleur de la tâche qui devrait être accomplie dans les quatre ou cinq années à venir montre que les structures actuelles doivent se développer ou que de nouvelles structures doivent être mises en place.

Au niveau "technicien et technicien supérieur", la tâche est peut-être encore plus importante en volume, mais elle l'est beaucoup moins en variété. C'est au niveau de la conception des ouvrages et machines que les idées nouvelles jaillissent le plus abondamment : *Concorde* est un avion inaugurant un domaine technique entièrement nouveau[1] et pourtant il est construit avec des procédés et outils classiques[2].

Cependant, même au niveau "technicien", il faut introduire les notions qui n'étaient pas enseignées voilà cinq ou dix ans. Ce sont toutes celles qui touchent à l'informatique, aux machines à commande numérique et à la gestion. Les procédés d'usinage par électro-érosion sont connus depuis quinze ou vingt ans et ils sont entrés très progressivement dans l'atelier ; les procédés de soudure évoluent très lentement : l'apport annuel des connaissances techniques nouvelles est relativement faible, si bien que le recyclage ne touche pas un très grand nombre de techniques. L'enseignement est donc plus facile à organiser puisque l'éventail est plus fermé et les "classes" plus nombreuses.

Mais dans les deux cas il est nécessaire que la liaison entre l'industrie et les centres de formation postscolaire soit bien assurée, puisque l'objectif premier est d'aboutir à un meilleur développement de l'industrie du pays.

On pourrait voir dans ce processus une tendance à accélérer aveuglément le progrès, alors que d'un point de vue philosophique il ne semble pas évident que ce seul objectif soit humainement valable. . .

Si, comme nous le souhaitons, les centres de formation postscolaire ne revêtent pas un caractère purement commercial, si, par conséquent, ils sont à caractère non lucratif, et étroitement liés à une école, on peut espérer que les gouvernements ou les fondations qui apporteront leur concours à ces centres auront la sagesse d'inciter leurs auditeurs à réfléchir sur certains problèmes tels que ceux des pollutions et des nuisances.

Il est difficile de donner dans l'enseignement de base (cycle scolaire de l'adolescence) des données précises sur la pollution et sur les nuisances. Avant de

1. On compare souvent *Concorde* aux avions militaires américains supersoniques tels que le B70. Mais on oublie deux particularités fondamentales: *a*) l'habitacle à "conditionner" n'est pas seulement celui de l'équipage, mais c'est aussi celui de la cabine des 128 passagers; or la température de surface de l'avion sera voisine de 100 °C; b) Un avion militaire a comme "charge payante" (! ) des "objets" à forte densité; l'avion est donc petit et par suite rigide; *Concorde* est calculé en avion souple (déformable).
2. C'est là un exemple typique de la notion de "système": c'est, en effet, l'intégration des différents composants, chacun à la pointe de la technique correspondante, qui constitue la véritable difficulté de la réalisation d'un projet tel que *Concorde.* C'est aussi la raison pour laquelle l'enseignement de recyclage doit exceller dans ce domaine.

limiter, de restreindre, il faut connaître les phénomènes physiques fondamentaux (il faut, cependant, sans aucun doute, mentionner à l'occasion de l'étude de chaque phénomène les risques de nuisance qu'il comporte). Par contre, dès qu'une certaine expérience est acquise et que les responsabilités de l'individu croissent, il faut attirer son attention sur les conséquences néfastes des objets qu'il imagine.

Enfin, nous mentionnerons, sans nous appesantir, les immenses possibilités qu'offre l'enseignement permanent à quiconque désire "faire autre chose", "explorer un nouveau domaine", bref "s'aérer" en rompant avec la monotonie d'une tâche sans cesse exécutée. Il est bon, certes, que chaque individu ait à cœur de mener à bien les tâches qui lui sont confiées, d'approfondir les sujets qu'il traite ; mais la nature humaine aime en général le changement et il ne faut pas croire que l'enseignement permanent est fait pour des Pasteur ou des Einstein.

L'enseignement de recyclage ne doit donc pas être seulement un facteur d'accélération du progrès technique ; il doit également : *a*) donner à chaque individu qui le désire la possibilité d'accroître ses connaissances, à titre personnel éventuellement ; *b*) éveiller le désir de ne faire progresser l'économie d'un pays qu'en parfaite harmonie avec le bien-être de ses habitants ; *c*) donner aux individus la possibilité de changer de domaine pour rompre la monotonie d'un métier qu'on exercerait sa vie durant.

Quant aux nouvelles structures que nous évoquions plus haut, elles peuvent rester dans le cadre des écoles si celles-ci veulent bien admettre que, d'ici dix ou quinze ans, leur mission aura basculé : au lieu d'avoir pour activité principale la formation *ab initio* des jeunes et accessoirement celle de recycler des ingénieurs, elles auront pour population majoritaire des ingénieurs en recyclage et, noyés dans cette masse, une petite minorité d'élèves-ingénieurs.

C'est une question d'organisation interne. Les élèves n'occupent les laboratoires que peu d'heures par an ; les auxiliaires audio-visuel permettent de répéter des séances hors la présence du professeur (mais, bien entendu, avec les assistants).

Nous estimons que Sup'Aéro pourrait accueillir, dans l'avenir immédiat, sans construction nouvelle et avec peu d'investissements techniques supplémentaires (de l'ordre de 10 à 15 %) une population résidente d'environ 250 stagiaires répartis en 10 stages simultanés, à partir du 1er septembre jusqu'au 15 juillet. (La durée moyenne des stages étant de deux semaines, il s'agirait d'une population constamment renouvelée.) Compte tenu des vacances scolaires des élèves, on voit que le nombre des journées-stagiaires (50 000) serait déjà plus important que le nombre actuel des journées-élèves (45 000).

Il semble que la plupart des écoles soient dans la même situation que Sup'Aéro.

Les structures à remodeler ou à innover ne doivent donc pas être un écueil à la mise en place de l'enseignement permanent – pour les ingénieurs et pour les autres – dans les meilleurs délais puisque cette activité constitue indéniablement un facteur de progrès social.

# Annexe

## ANALYSE SOMMAIRE DES TEXTES RELATIFS A LA FORMATION CONTINUE (PLUS PARTICULIEREMENT POUR CE QUI CONCERNE LA FORMATION DES INGENIEURS ET ORDRES SUPERIEURS)[1]

On trouve dans les textes des expressions différentes qui se recouvrent plus ou moins complètement: formation ou éducation permanente ou continue, promotion sociale, perfectionnement, recyclage.

Ici, le terme "formation continue" sera employé au sens de "formation professionnelle continue", englobant tout ce qui se situe après la formation initiale.

Sans remonter aux protocoles de Grenelle (1968) ou même à des textes antérieurs, on peut dire que la formation continue en France a été : *a*) définie par les accords nationaux interprofessionnels de 1969, 1970, 1971 ; *b*) organisée par les lois du 16 juillet 1971 et les textes d'application (en cours de parution).

### *Accord du 9 juillet 1970* sur la formation et le perfectionnement professionnels (suite à un accord interprofessionnel du 10 février 1969 sur la sécurité de l'emploi)

L'accord du 9 juillet 1970, basé sur la certitude que "l'enseignement et la formation sont facteurs du progrès social, du développement technique et de la croissance économique", avait pour but de définir "une politique propre à donner à la formation et au perfectionnement professionnels les moyens de leur rapide développement", pour objectif "de répondre aux besoins des entreprises et de permettre aux individus d'y trouver la satisfaction de leurs aspirations" et pour résultat de "situer la France tout à fait à l'avant garde dans ce domaine", en apportant aux salariés des "garanties qui n'existent encore dans aucun autres pays européen". Il a été signé au nom d'environ 10 millions de salariés.

L'accord comprend deux parties principales :

#### *1. Formation générale des jeunes*

*a*) Fréquentation obligatoire (sans perte de salaire) des cours professionnels jusqu'à 18 ans (19 ans pour les apprentis) dans la limite de 320 heures par an et de 8 heures par semaine ; ***b***) apprentissage, indemnisation, placement par les centres professionnels ; *c*) contrôle par des représentants des travailleurs ou des commissions partaires.

1. Faite par M. Clérouin, adjoint au directeur des études de l'ENSAE.

#### *2. Formation des adultes*

*a*) Adultes faisant l'objet d'un licenciement collectif (autorisation d'absence pour stage ou session de formation) ; *b*) adultes en activité (autorisation d'absence pour formation).

L'accord prévoit de plus la création, au niveau national interprofessionnel, d'un comité paritaire pour la formation et le perfectionnement.

### *Avenant du 30 avril 1971 à l'accord du 9 juillet 1970, concernant plus spécialement les cadres*

Il a été établi pour tenir compte des "responsabilités particulières des cadres" des "aspects spécifiques des fonctions qu'ils occupent" et de la "nature des travaux qu'ils conduisent", ainsi que du fait que la "qualité" de leurs actions "est un facteur essentiel des progrès techniques et technologiques" et de l'"efficacité des entreprises".

Il s'adresse aux ingénieurs et cadres et peut s'appliquer aux agents de maîtrise et techniciens (après accord paritaire au niveau des professions). Il nous a pour domaine "l'acquisition, l'entretien, la mise à jour et l'approfondissement des connaissances de base, ainsi que l'élargissement de la formation générale".

Les dispositions de l'accord du 9 juillet 1970 sont applicables aux cadres avec des variantes peu importantes.

*Lois du 16 juillet 1971*

*a*) Loi 71.575 portant organisation de la formation professionnelle continue dans le cadre de l'éducation permanente ; *b*) loi 71.576 relative à l'apprentissage ; *c*) loi 71.577 d'orientation sur l'enseignement technologique ; *d*) loi 71.578 sur la participation des employeurs au financement des premières formations technologiques et professionnelles.

Les lois 576, 577 et 578 présentent un intérêt réduit pour la formation continue. A noter cependant les points suivants :

*Loi 576*
L'apprentissage dure en principe 2 ans. Des centres de formation d'apprentis sont créés. L'apprenti est engagé entre 16 et 20 ans. Son contrat prévoit qu'il sera instruit, partie dans l'entreprise, partie dans un centre de formation, et que cet enseignement comptera comme horaire de travail. Date de mise en application : contrats établis à compter du 1er juillet 1972.

*Loi 577*
"Les enseignements technologiques sont constitués par l'ensemble des moyens destinés à assurer la formation professionnelle initiale et la formation continue. Ils assurent "un ensemble de formations pouvant s'étendre de la troisième année du cycle moyen jusqu'à l'enseignement supérieur inclus" et comportent "un stage d'initiation ou d'application en milieu professionnel". "Ce stage fera l'objet d'un contrat entre l'établissement d'enseignement et l'entreprise."

*Article 9*. "Au plus haut niveau de l'enseignement et de la recherche, les disciplines technologiques sont consacrées par des diplômes délivrés dans le cadre de la loi d'orientation de l'enseignement supérieur du 12 novembre 1968 ou de la loi du 10 juillet 1934 relative à la délivrance du diplôme d'ingénieur."

*Loi 578*
Taxe d'apprentissage ou "premières formations technologiques et professionnelles", avant l'entrée dans la vie active.

La taxe est ramenée de 0,6 à 0,5 %

Elle s'applique toujours aux ingénieurs ou cadres supérieurs. Le pourcentage de la taxe affecté n'est pas connu.

*Loi 575*
Elle est, pour ce qui nous concerne, la plus importante. Elle reprend et codifie les principales dispositions de l'accord du 9 juillet 1970.

Son esprit est résumé par l'article premier : "... la formation professionnelle permanente constitue une obligation nationale. Elle comporte une formation initiale et des formations ultérieures destinées aux adultes et aux jeunes déjà engagés dans la vie active ou qui s'y engagent. Ces formations ultérieures constituent la formation professionnelle continue.

"La formation professionnelle continue fait partie de l'éducation permanente. Elle a pour objet de permettre l'adaptation des travailleurs au changement des techniques et des conditions de travail, de favoriser leur promotion sociale par l'accès aux différents niveaux de la culture et de la qualification professionnelle et leur contribution au développement culturel, économique et social.

"L'Etat, les collectivités locales, les établissements publics, les établissements d'enseignement publics et privés, les associations les organisations professionnelles, syndicales et familiales, ainsi que les entreprises, concourent à l'assurer."

La loi comprend 8 titres, dont les dispositions concernant la formation continue sont résumées sommairement ci-après :

*Titre premier. Institutions de la formation professionnelle. a*) Auprès du premier ministre, un comité interministériel dont le ministre de l'éducation nationale est vice-président, un groupe permanent de hauts fonctionnaires dont le président est nommé par le premier ministre, un Conseil national de la formation professionnelle, de la promotion sociale et de l'emploi (représentants des pouvoirs publics et des organisations professionnelles et syndicales) ; *b*) à l'échelon régional et départemental, des comités de la formation professionnelle, de la promotion sociale et de l'emploi.

*Titre 2. Conventions de formation professionnelle.* Les actions de formation professionnelle font l'objet de conventions déterminant la nature, l'objet, la durée, les effectifs des stages, les moyens pédagogiques et techniques mis en œuvre, les conditions de prise en charge des frais. Les conventions sont à passer entre les organismes demandeurs (entreprises, associations, établissements, organismes, collectivités, organisations, etc.) et les centres de formation.

*Titre 3. Congé de formation. a*) Durée d'un an à plein temps ou de 1 200 heures réparties sur un cycle pédagogiques d'instruction, davantage en cas de stage de "promotion professionnelle" ; *b*) Conditions exigées des bénéficiaires : avoir 2 années de présence dans l'entreprise, avoir obtenu depuis plus de 3 ans un diplôme d'enseignement professoral ; *c*) Nombre d'absences simultanées : le nombre des heures d'absences ne doit pas excéder 2 % du nombre d'heures effectués par l'entreprise ; *d*) La rémunération des stagiaires peut être assurée par l'entreprise avec aide de l'Etat.

*Titre 4. Aide de l'Etat.* L'Etat concourt au financement en participant aux dépenses de fonctionnement des stages, de construction et d'équipement des centres. Les crédits sont inscrits au budget des ministères intéressés ou à celui du premier ministre. La contribution peut porter sur quatre sortes de stages : stage de "conversion" et de "prévention" (mieux orienter les bénéficiaires), stage d'"adaptation" à un premier ou à un nouvel emploi, stage de "promotion professionnelle", stage d'entretien et de perfectionnement des connaissances.

*Titre 5. Participation des employeurs.* Début en 1972 : 0,8 % au moins des salaires, jusqu'à 2 % en 1970. Les employeurs peuvent au choix : *a*) organiser eux-mêmes la formation continue ou passer des conventions avec les centres de formation : *b*) effectuer (dans la limite de 10 %) des versements à des organismes "soit agréés sur le plan national en raison de l'intérêt que présente leur action pour la formation professionnelle continue, soit menant des actions dont l'intérêt sur le plan régional a été reconnu par le préfet de région sur proposition du comité régional" ; *c*) participer à la constitution des fonds d'"assurance-formation".

*Titre 6. Aides financières accordées aux stagiaires.* Ce titre, très long, analyse les conditions de l'aide apportée aux stagiaires par "l'Etat, les employeurs, les travailleurs et les organismes chargés du service d'allocation d'assurance aux travailleurs sans emploi".

*Titre 7.* Il concerne les dispositions relatives aux agents de l'Etat et prévoit que des décrets fixeront la politique dans ce domaine (organisations, conditions d'accès, rémunération). Il est prévu que "des instituts régionaux d'administration à créer par décret pour le recrutement et la formation de certains corps de catégorie A pourront prêter leur concours à la formation continue des agents de l'Etat".

*Titre 8. Dispositions diverses.* Elles concernent les professions agricoles, la formation continue des chefs d'entreprise à laquelle l'Etat contribuera), la création d'une "fondation nationale pour l'enseignement de la gestion des entreprises" au profit des "cadres occupant des emplois de responsabilité".

*En résumé*

1. Toute entreprise de plus de 10 ouvriers doit consacrer 0,8 % (2 % en 1976) des salaires bruts de toute nature à la formation continue. Elle peut : soit organiser elle-même la formation continue, la financer à tous les niveaux (salaires et charges sociales des travailleurs et des enseignants, frais d'équipement et de fonctionnement des centres) ; soit avoir recours à une aide extérieure (conventions annuelles ou pluri-annuelles avec des organismes de formation, versement de 10 % à des organismes agréés) ; soit constituer un fonds d'"assurance-formation" au profit d'un organisme distinct de l'entreprise mais dont elle pourra conserver la gestion.

2. Tout salarié peut : soit bénéficier de cette formation continue ; soit demander un "congé de formation", sous les réserves indiquées plus haut ou un "congé pour aller enseigner" (il existe aussi des congés de "conversion" en cas de licenciement collectif et des "congés spéciaux de 100 heures au maximum pour les jeunes travailleurs de moins de 20 ans ayant moins de 2 ans de présence dans l'entreprise et ne possédant pas de diplôme professionnel).

3. La rémunération est assurée : par l'entreprise pour les salariés qu'elle envoie en stages, ou qui sont en congé de conversion ou de formation (pendant 160 heures dans ce cas) ; par une aide de l'Etat à l'entreprise ou directement au stagiaire sous réserve qu'il s'agisse d'une des catégories de stage prévue par la loi. L'aide de l'Etat varie de 30 % (stage d'entretien) à 70 % (stage de conversion).

# The continuing education of the engineer (summary)

Marc Pelegrin

Continuing education may be regarded as an important factor in national economic development, and also may be seen as a means of giving the opportunity of obtaining additional knowledge to individuals who by accident or social background have been denied full initial education.

In a field such as aeronautics the progress made in technology in the last several decades has been such that updating of knowledge has been indispensable to all practising engineers. It is in this context that the Ecole Nationale Supérieure de l'Aéronautique et de l'Espace has established several distinct continuing education programmes, acting in collaboration with the concerned industrial establishments and a specially created private organization of friends of the school. The choice of topics is based on the immediate needs of industry and on estimates of the long-term evolution of technology.

The growth of continuing education in France has been facilitated in recent years by a series of government measures culminating in the law of 16 July 1971 that established the framework of continuing education in France and the means by which it is financed.

# Demands of society with regard to the continuing education of engineers

H. Hahn

'Education is by far the largest single economic effort' writes L. V. Berkner [1][1] in a fictitious report for the year 1984. While Berkner predicts for 1984 that primarily education at the doctoral and post-doctoral level will be 'the bottleneck of economic development', today education at all levels presents the most serious problems. Progress can be defined exclusively as economic growth (Berkner) or understood as a general improvement of the conditions of life experienced by the majority of mankind. This progress will be governed by the quality and extent of education which mankind has attained. In these days of concern for our environment and its possible deterioration, one can state the relationship between education and progress even more pointedly : the survival of mankind is endangered by all the possibilities of self-destruction that we have developed through our knowledge. To overcome this threat we need far-reaching, carefully planned lifelong education. One can demonstrate the beneficial effects of imparting meaningful information to all segments of society, whether they are responsible for progress or simply experience its positive and negative consequences, especially in the area of progress and environmental protection or deterioration.

In view of the importance of education as the bottleneck of progress, the question of 'continuing education for engineers' will be discussed in the context of national economic and social planning. The arguments used in the discussion are based upon the author's experience as a university teacher. Therefore this contribution is by no means the result of statistically designed investigations, nor does it claim to represent a complete survey of the whole field of continuing education in the engineering disciplines. It is rather the outflow of positive and negative experiences in training and teaching engineering students and includes impressions gained in many discussions between 'men from theory and men for practice'.

Society at large should have a paramount interest in continuing education in general, and in such enterprises in the realm of engineering in particular. Among the reasons for this ever-growing interest, the following are of great importance:

1. Education today represents a large public investment. Only continued education will ensure the continuing activity of such investments.
2. The time allowance for basic professional education in engineering subjects is reduced today. This development, which is caused by technical, economic,

1. Figures in brackets refer to references at the end of this article.

sociological and other phenomena, may be compensated by intensified 'post-graduation' education.

3. Continued education establishes a feed-back circuit between the places of teaching and research on one hand and practice on the other. In this way the always present and sometimes dangerous gulf between these two vital estates of society may be bridged.
4. The increasing number of decisions to be made regarding technological and scientific questions requires an effective dissemination of information if the decision-making is to be democratic and responsible. The establishment of facilities for continuing education for even one segment of society will be helpful for the education of society in general. Thus, continued engineering education may be of direct benefit to society at large.

Recognizing these circumstances, society will have to ensure that facilities for continued education are created and supported, or, where they are already in existence, that they are well staffed and widely used.

## EDUCATION—A PUBLIC INVESTMENT WHICH IS ACTIVATED BY CONTINUED EDUCATION

Today we are witnessing a trend away from privately financed education and towards institutions which derive their funds largely from official sources. One may look at the American scene, for instance, where well-endowed private universities begin to show signs of financial difficulties as compared with the situation of State-funded universities, or at the situation in the Federal Republic of Germany where in most instances all fees for education at all levels have been abolished. Everywhere the increasing costs of education tend to make the private financing of institutes of education difficult or impossible.

It is not only the growing difficulty of financing education which call for a public interest in educational matters. Systems built on private initiative have, whether consciously or by sheer coincidence, developed only part of the available resources in manpower. Yet it becomes more and more important to give equal possibilities for education to all sectors of society. This is not only a necessity from the social and political viewpoints but is also dictated by the need for well-trained manpower in all fields. The extent of this training and the number of trained specialists must keep pace with the rapid developments of progress. It has been amply demonstrated that the so-called environmental crisis, triggered by an incomplete understanding of environmental facts in the producing and consuming segments of society, is to a large degree a manpower crisis [2]. Without too much imagination one can predict that the potential strength and power of a nation will be measured less and less by its physical resources and material possessions or by the amassment of arms, and increasingly on the basis of its intellectual potential [3]. In making this statement, it is recognized that competition of all sorts at this level will be detrimental to mankind; only international co-operation will ensure beneficial developments from an attempt to improve the educational level of the majority of mankind [4]. Thanks to mechanization, automatization and the increased use of computers in all fields, we are witnessing a development that frees man more and more from the hardship of physical work and allows a reduction of working time in almost all professions. It is to be expected that a large part of this newly gained leisure time will be spent not only in traditional recreation activities but also to the end of gaining better information and continued education. The very 'progress' that demands a better understanding of the possibilities and limitations of life on this planet, gives us more freedom to acquire such knowledge.

If official agencies are to be engaged in the politics of education owing to their financial obligations, they can be expected to demand a fair share in the process of policy-making. This will in many cases lead to difficulties and confrontations between the power-wielding delegate of society and the specialist who alone feels responsible for all aspects of the subject matter. It is illusionary to expect that such differences of opinion will ever vanish or can be eliminated by regulations. However, the orientation towards one common goal and the discipline demanded by objectivity will bring out the necessary co-operation. Whatever the points of discussion or disagreement may be, one positive aspect of a more co-ordinated and controlled educational policy is likely to emerge, namely a reasonable and realistic distribution of the available manpower resources to different fields. Today, at a time of rapid shifts in the importance and relevance of different disciplines, we still permit decisions concerning the orientation towards one or another field to be made exclusively by the individual and frequently on the basis of irrational arguments. It appears that, in particular, scientific and engineering disciplines suffer from such anti-rational behaviour [5].

Yet it is precisely in these disciplines that investments are very high and returns are small. The costs of building, maintaining and operating educational facilities for the training of scientists and engineers are rising constantly. It is not only the increasing complexity of the necessary experimental equipment but also the long-term investments that must be made in educating good and up-to-date teachers, which cause a more than proportional rise in expense in this area. It is against this background that one must view the growing danger of obsolescence in scientific and engineering subjects. The rapid loss in relevance of acquired knowledge, expressed as turn-over rate or half-life of information, is an intrinsic characteristic of science. It is caused not only by great revolutions in scientific disciplines such as the breaks in continuity of scientific thought that accompany the evolution of nuclear science and the revelation that matter can be converted into energy and vice versa. There are also day-to-day advances which cause irreversible changes in scientific thinking. The half-life of knowledge is frighteningly short [6]. Yet it must be expected to decrease in the days to come.

There are still other, less obvious phenomena, that explain the considerably smaller returns from investments in the education of scientists and engineers as compared with those in other areas. To mention only one, engineering practice now tends to solve problems by team-work. The amount of information and knowledge handled in the course of a project is too much and of too great a diversity to be coped with adequately by one type of engineer. Specialization has penetrated engineering practice to a larger degree than in many other fields. The contribution of an individual, the specialist, to the solution of a particular problem is therefore comparatively small. And the consequent intensive specialization of all members of the engineering profession leads them into one-way streets from which they can return only at great expense; moreover, the one-way street very often leads to an unexpected dead-end, owing to the fast turn-over rate of knowledge.

In many instances it is not just one team that works on the solution of a given problem; it has become customary in some branches of engineering to make use of the driving force of competition by directing several competing teams to work on the same question. Here the actual advance due to any one individual is even smaller. The return from investments made in the education of engineers is thus reduced once again, that is if one defines return by the direct benefit that the work of any one trained engineer brings to society. In this connexion one might point out that society is much more aware of advances or stagnation in the field of engineering than in many other fields and therefore keeps a close count of the so-called returns.

Finally, one must mention the considerable brain drain suffered by the engineering community due to the custom of recruiting experienced and professionally somewhat obsolete engineers for higher management positions. However, it happens that the managerial knowledge required in the new position is self-taught. Therefore where the individual's engineering accomplishments are reduced drastically as a result of the change of position, his achievements in the new role may not on their own justify the large investment in his professional education.

The small returns from the large public investments that are necessary for the education of scientists and engineers have triggered many investigations primarly concerned with methods of combating obsolescence and professional brain drain in the fields of engineering. It has been argued that the teaching of the foundations of engineering work, i.e. imparting a knowledge of the underlying scientific theories, gives the practising engineer the greatest flexibility and adaptability with respect to the acquisition of new skills. Yet it is not only the tools of the engineer, that is the application of basic knowledge, that change. Thus one might well ask in which sector of engineering education and practice the phenomena of obsolescence are most severely noticed: in the advances of basic sciences, in the ensuing shifts in the application of new knowledge, or in the renewal of methods of analysis and problem-solving used in the realm of engineering? On which of these sectors should the initial phase of education concentrate and where will subsequent continuing education be most successful? Or might it be possible and also expedient to renew and reactivate in every phase of continuing education all three sectors of engineering know-how? These questions must inevitably conjure up some fear of a future that offers the grim prospect of a rapidly growing rate of professional obsolescence. It is not surprising that concerned individuals ask about a possible control, or even reduction, of the rate of renewal of knowledge and ideas.

On the basis of experience gained in university teaching one is tempted to emphasize the teaching of study methods to the young engineer, that is, developing an ability to acquire diversified and detailed knowledge by oneself through working with real-world problems. Thus one would expect continuing education to produce the best results when it concentrates on reactivating the ability to study effectively and critically and to transfer the acquired information into practice. To express this in more practical terms, the first phase of professional education should – using practical examples – concentrate on teaching problem-analysis and problem-solving, whereas continuing education, that is, the following phases of professional training, should concentrate on teaching the way to acquire and apply new basic knowledge. The step from the incorporation of new basic knowledge to its practical application is not necessarily an easy one, but is more within the power of an alert engineer in practice than the following-up of new developments in basic sciences, the selection of relevant information and the preparation for a successful assimilation of this new knowledge. Considerations with respect to existing teaching methods and teaching staffs and possibilities as well as limitations in establishing new facilities will without doubt also influence the decision about the content of continuing education.

Professional journals and their documentation provide valuable support to efforts towards the establishment of continuing education. It is therefore of great importance to recommend this element of information, communication and discussion to the young student as well as to the experienced practitioner. Nevertheless, the content and form of professional publications might be better adapted to the purpose of continuing education. This is not meant as a criticism of their scientific or technical quality, but rather a suggestion to include pedagogical considerations in the writing or reporting of scientific and technical advances.

While there may be general agreement on the necessity of continuing education and its motivation as well as its realization, a large number of detailed questions remain to be answered before any such programme can operate sucessfully. Most of these questions will be asked by society at large, because society eventually finances these efforts and expects to reap the benefits. To a large extent these will be questions that cannot be answered on the basis of experience with teaching or with the subject matter to be taught. The questions may range from the problem of determining quantitatively the state of obsolescence all the way to the problem of checking the efficiency of the established educational system. In general, the ensuing discussions will centre around two key issues: first, the role to be played by official agencies in policy-making within the realm of continued education, and second the effects of such evolution upon the individual concerned and his place in the various social settings he belongs to. If one discusses questions as to the extent and form of continuing education one will always have to take into consideration the ideas, wishes or demands of the financing agency, be it the government or industry. This same supporting group will also want to reserve a right to decide who takes part in such educational enterprises, how its effects are to be measured and effectivity is to be maintained, and, if necessary, where shifts from one discipline to another must be made.

While such decisions significantly affect the individual who seeks further education, there are other questions that are of even greater importance to him. Can he, the practising engineer, be forced to participate in this phase of his professional training? Such pressure could be exerted on him in order to keep active a financial investment to which he contributed relatively little. But, what will become of the well-established, and presumably well-functioning professional hierarchies? These are based more on practical experience than on active and relevant knowledge. The established hierarchy may break when young and old engineers compete directly in acquiring skills and know-how. More important, it has been possible up till now to map out for oneself a development and possibly a career on the basis of one's own initiatives, involvement and changing capacities. Now it must be expected that the individual engineer will be directed horizontally or vertically within his professional hierarchy as a result of decisions made by somewhat remote agencies. Such decisions will be made on the grounds of varying manpower needs in different areas. In the future, the problems at hand will determine the individual's assignment and the qualification he must acquire; traditionally gained qualifications will no longer guide the selection of tasks on which an individual will have to work.

Finally, it may be necessary to discuss the actual set-up of continuing education. Is it meaningful to integrate it fully into existing educational facilities and study plans or will it be better to develop parallel but independent facilities and study models? Whatever the final form and content of continuing education for engineers may be, it is mandatory to see the urgent need for such educational institutes as a means of enlarging the activity of large public investments. Thus, difficulties in deciding the above-mentioned questions should by no means be allowed to delay action.

## CONTINUING EDUCATION CAN COMPENSATE THE REDUCTION IN TIME AVAILABLE FOR BASIC PROFESSIONAL SCHOOLING

The more and more discernable tendency towards a shortening of study time in the first phase of professional education documents is an irreversible development. There are many, in particular amongst the specialists in the educational field, who decry this phenomenon. Yet it seems more useful to devise some means to compensate for the loss of study time. One of these possibilities may be found in continuing education.

The reasons for the significant reduction in time allotted to the education of engineers are manifold. The severe crisis in education, which is coupled by some sort of feed-back relationship to crises and difficulties in other areas of life, seems to be foremost a manpower crisis, together with a general lack of material resources. Thus a stricter regulation of the educational process, especially a more disciplined and intensified curriculum, absorbed in a shorter period of time, is expected to overcome the crisis and create an escape from the feed-back relationship.

However, while study time is shrinking, the amount of knowledge to be transferred is growing. This generally acknowledged phenomenon was mentioned above. It is not only the amount of scientific or technical knowledge needed for solving a problem which grows rapidly. The engineering curriculum is burdened more and more by the so-called 'non-technical' understanding and knowledge demanded by an alarmed society from the engineer. This, too, effectively reduces the time allowed for the study of engineering skills [7].

The economic structure that our society has created leads to an over-growing demand for well-trained scientists and engineers, amongst other specialists. But from the moment of choosing an educational path with a view to entering a profession, the student is tempted by the generous offers, of industry for example, to speed up the slow and apparently unproductive process of professional training. Who would blame today's students for not subjecting themselves to the trials of truly academic and self-disciplined studies, when the promises of our expanding economy constantly lure them away from non-productive, intellectual achievements to constructive problem-solving?

There is another stimulus to speedy progress through one's course of study, namely the changing age structure of many professional hierarchies which leads to the appintment of very young people to leading positions. To avoid the risk of missing the best opportunities, the student must assimilate the necessary and irksome studies with maximum speed and economic effectiveness.

Independently of the reduction in time allotted to the first phase of professional education and training, there is a trend towards more practice-oriented teaching and research which may solve some of the above-mentioned problems. It means that both segments of the engineering community, the teaching and research engineer and the practising engineer, which have been separated at times, will move closer together.

The engineer's education in the past could, with noteworthy exceptions, be paraphrased as an education for the 'blue-print type' of engineer. That type of professional education was essentially a training in the solution of standard problems that were similar in concept to real problems. Such training was meant to enable the graduate student to apply the recipes developed or presented during his schooling to any real-world problem. But the growing size and complexity of today's engineering tasks require 'tailor-made' solutions. The repeated use of a blue-print that has proven successful in other cases becomes increasingly difficult. In passing, it should be mentioned that it is the relatively new engineering concept, namely to consider projects not in isolation, but in the context of their immediate or more distant environment, that demands specific solutions and therefore a differently trained engineer. Thus, today's teaching tends to concentrate on analysing and solving difficult real-life problems in such a manner that the student learns approaches or methods of analysis and problem-solving rather than ready-made recipes.

The mounting difficulties associated with the solution of engineering problems have also led to some pooling of manpower resources. A rising number of large projects are worked on jointly by men from practice and men from teaching and research institutes.

There are other reasons for such a moving together of practice and theory which are of specific importance to the university itself. It may appear difficult at times to evoke enthusiasm in a student for his studies. But one of the ways to generate and maintain the necessary motivation is to introduce real-world problems into a curriculum that is filled with the study of basic sciences and the application of such knowledge.

Furthermore, the great demand for scientists and especially engineers leads industry, or more generally, practice, to ask for practically and realistically trained students who can apply themselves to productive work without delay. The request for such an orientation towards reality and direct problem-solving does not necessarily mean a return to the former blue-print type of engineer.

The greater interest of universities in the questions and problems posed by practice is not the only possible way to compensate for the reduction in the time allowed for basic professional education. The increasing importance of team-work and its introduction into the realm of the university has two significant consequences: first, a concentration upon problems that justify the expense of team-work, and, second, co-operation between experienced specialists rather than between students of one particular discipline. Today, team-work is a word which is as frequently abused as the term 'environment'. There is nothing fashionable about team-work, at least not to those who are familiar with the inherent problems of bringing together differently trained and differently functioning individualists. This method of solving complex problems has always existed; it has gained importance, however, in recent times, as a necessary compensation to the horizontal specialization in all disciplines. The process of 'atomization' [8] which manifests itself in many areas of life will accelerate and become more important. And along with this separating and splitting-up of subjects and disciplines, compensatory measures will become more and more necessary and also increasingly successful.

One should not make the mistake of thinking that at earlier times, when specialization had not gone so far, the competent and trained engineer, who possibly has a representative knowledge of the whole field, applied his comprehensive knowledge at all times to each task. Then, as today, the engineering discipline had its outstanding geniuses, who were but few in the multitude of average practitioners. Today we are more aware of insufficiency and mediocrity and we want to ensure the optimal solution of a large number of projects. We hope to find the right solution with greater certainty than would be obtained by relying on the random appearance and inspiration of an outstanding leader in his profession. Team-work appears to be one of the few methods which give some guaranty that we will go beyond mediocrity in our work.

We have seen that the growing emphasis on team-work, even in the first phase of basic professional education, leads to a closer connexion between practice and the institutions of learning. This development prepares the way for continuing education. From a different viewpoint one could also argue at the introduction of continuing education into the realm of engineering will facilitate the necessary inclusion of team-work approaches in study curricula. Everyone will agree that actual team-work cannot be applied to 'standard school problems' which are frequently of a hypothetical nature. To show the very merits of this method, as well as to demonstrate its possible difficulties, it must be applied to real, complex engineering tasks. Moreover, most teams consist of specialists from different disciplines and it is this combination of different branches of knowledge and the co-operation between experienced professional workers that gives team-work its practical value. But if a team is composed solely of students from one discipline, it is not likely to yield 'the results expected of team-work'. Therefore the training of engineering students to work in interdisciplinary teams can be successful only if men from practice contribute their experience and active co-operation to the teaching of young engineers.

These developments, be it the reduced time of basic professional education, or the closer contact between teaching and practice, will give greater significance to 'on-the-job' study and training within the formal curriculum. Up till now, it has been customary to plan engineering curricula with the main object of transferring basic knowledge, and to leave the step of converting well-trained but inexperienced students into useful partners completely to practice. Now both these phases of engineering must be planned together. And if this is done in conjunction with the planning of continuing education for practising engineers, all participants will profit to the maximum.

Such a combination seems to guarantee a meaningful mixture of 'theory' and 'practice' in the process of training new engineers. In planning engineering curricula we have frequently experienced the full swing of the pendulum from one extreme to the other – at one time a complete orientation to the needs of practice, and at other times a short-sighted worship of basic sciences and so-called theory. Now the establishment of continuing education in conjunction with the first phase of engineering education makes it possible to dampen the swings of the pendulum and ensure a meaningful balance between the extremes.

The practical plans and programmes adopted for continuing education will influence the form of co-operation between the university and the profession. All plans, however, should be guided by one conclusion: engineering education is never completed, either at the university level, or after many years of practical experience. If the engineer of today is to justify the hopes put into his work he must think of himself as an eternal student, always ready to exchange obsolete information for new practical or theoretical knowledge.

## CONTINUING EDUCATION ESTABLISHES A FEED-BACK RELATIONSHIP BETWEEN TEACHING AND RESEARCH ON THE ONE HAND AND PRACTICE ON THE OTHER

The moving together of the theory-oriented and the practice-minded segments of the same field means, on the one hand, a joining of efforts and a pooling of resources. Yet there is another very useful function that continuing education might possibly serve: that is, the mutual inspiration and fertilization of work in both areas. This again is an aspect of great interest to society, a society that has suffered from the isolation of these two segments.

It was mentioned above that the planning of engineering curricula has frequently suffered from concentration on extreme and varied interpretations of the engineer's role and the demands that society makes upon him. If the university shares the privilege of setting up curricula with other interested and qualified men, the responsible participation of experienced alumni may prevent excessive fluctuations. In particular, this may help to provide conclusive and convincing answers to the question: Where can one omit topics from the curriculum, and where is the emphasis to be laid, in order to comply with the demands for intensification of study? Presumably those who deal with the application of engineering knowledge will be able to identify what is most important to the beginner: for example a sound foundation in basic sciences, a knowledge of the application, or a good training in engineering methods. Such a joint council might also find more easily a solution to the problem of 'non-technical' education in engineering disciplines: where and to what degree is it meaningful if at all ?

In trying to solve such questions by co-operation and mutual agreement, members of the teaching segment and members of the practising sector will eventually also clarify the question of length of studies. There must be a possible compromise between the urgent demands of practice for effectively and quickly trained engineers and the universities' desire to furnish students with a very sound education. A similar compromise must be found with respect to the degree of specialization that the graduating student should attain. Should he still be able to span the whole engineering discipline, or would it be better if he concentrated his final studies on one facet of the whole field, as practice will eventually require him to do?

Finally, such joint discussions on curricula between engineering teachers and engineers returned from practice will help to avoid another mistake which is sometimes made in designing new educational models. In some instances, the planning of a new curriculum is largely guided by an urge to perform experiments of a more pedagogical nature. The content of the new curriculum is consequently determined in part by the form of information transfer, rather than the material to be taught influencing the method of teaching. In other circumstances, the curriculum must be reformed in haste to compensate for a development that has already taken place. Such planning and acting under pressure could be avoided by joint discussions designed to prepare for events when they are still barely discernible on the horizon of future developments.

Teaching and research are closely linked to each other. If the establishment of continuing education is going to affect the development of new forms, contents and goals in the area of teaching, it will also influence the planning and realization of research. Not all the questions that can be asked today in the area of applied or basic research can or will be answered. Leaving aside the most important consideration of ethics for the moment, the shortage of research funds and man-power, etc., confines us to the investigation of but a few of the possible topics [9]. In the process of selection of so-called relevant research topics, arguments as to the direct and indirect benefits of new knowledge for society will attain increasing importance. The flow of information concerning the needs and wishes of society will be carried to a large extent by the practising engineer, who returns to institutes of teaching and research for continuing education [10].

For the above reasons, continuing education will include the participation of the returning engineer in research as well. While it is very important to renew and reactivate knowledge, it is equally beneficial to demonstrate to the practising engineer the advantages of the 'scientific method'. This was paraphrased earlier as the attempt to find the one and only specific, 'tailor-made solution' to a given problem by thorough analysis and skilful synthesis of different observations. In view of the general acceptance of the working methods of scientists, it seems reasonable to encourage the practising engineer in critical and analytical thinking and to make him apply conceptual schemes synthesized from experimental results, which is the basis of the notable advance in science.

By participating in research, the engineer who returns for continuing education will develop a better understanding of the value of new findings, thus increasing the impact of new knowledge. Today the lag between the time of development and the time of application of new knowledge, although decreasing, remains considerable. The reasons for the period of unproductivity of newly established knowledge are manifold. For one, the presentation of new knowledge, mostly in the form of written publications, is often too abstract. A second phenomenon that may explain the delay is the following: those few who have grasped the possibilities of new information are always confronted in practice with a hierarchy and professional bureaucracy of large inertia. It is frequently the young engineers, just entering the field of practice, who want to bring their new knowledge to bear with great enthusiasm. And in such cases the resistance and inertia of the established professionals is even greater, and thus reduces the impact of recent findings even more.

Last, but not least, one must mention the overburdening of many practising engineers with routine and emergency work. In fairness, it must be said that this leaves them very little time for new theories, for any playing with new ideas, even though the individual engineer might be willing to do so. Opinions expressed at short-term renewal and refresher courses for engineers frequently underline this statement. Consequently those who are willing and who have gained some influence in their area of engineering work cannot attend continuing-education courses for fear of causing serious harm to their present projects.

However, it is not only subconscious resentments and uncontrollable forces that make the introduction of new engineering concepts into practice difficult or impossible. Conscious prejudices also represent barriers to reform. Although educational patterns have changed, the above-mentioned blue-print of engineer is still in existence. The mediocre engineer will use over and over again, even in quite different situations, what has proven successful in one case. Why not build on one's good experience? And instead of going through the whole development of this one successful idea, one uses only the results of the considerations. Furthermore, the development, content and meaning of new knowledge are, for reasons discussed earlier, rather remote from this not uncommon type of engineer. There is outright suspicion, fear or even scorn for anything new. He does not see any reason to throw overboard ideas that seem to have worked so well until now, just because something new has been discovered.

If it could be proved that the new approach is better than the old, then he would consider the innovation. But unfortunately, the full range of consequences of an engineering project can only be observed after considerable periods of time.

Conscious and seemingly well-founded resentments of the kind described above often lead to a thorough and successful brain-washing of the young, enthusiastic student who is about to enter practice. Instead of the student's convincing his older colleague, the latter accomplishes an amazing task of rapid re-orientation, re-education and conversion owing to his 'better-founded' argumentation. This brain-washing must occur for two reasons: first, there is the weight and authority of the older and more experienced colleague, and second, it is a necessary defence mechanism of the older and possibly more vulnerable engineer. If he did not defend himself by such attacks, the danger of his influence being diminished or his authority partly undermined would be too great. Not only does he feel insecure, he has not even had a chance of catching up with the newcomer on all that recently developed know-how.

The problem of a disturbance of well-established and even functional hierarchies owing to the introduction of continuing education has been mentioned briefly. If the return to school of both young and old men from practice might upset such structures by opening the door to true competition of knowledge against knowledge and experience against experience, how much more will the unconditional admittance of young engineers to the group of experienced men upset the order? One may, however, expect that being exposed to continued education, the hesitant practitioner will overcome this barrier.

To conclude these remarks about the establishment of a feed-back circuit between teaching and research on the one hand and practice on the other through continuing education, one might express the hope that such developments will lead away from the one-way communication between teacher and student that still exists in many universities. Partly as a prolongation of the authority formerly exercised by teachers, partly because of the student's own inclination, insecurity and inertia, he is today in many instances still a mute receptacle and non-critical consumer of the predigested information given to him. Most of the students want no part in pedagogical experiments that could deter them from their goal of gaining as quickly as possible a basis for entrance into the professional world. It is of greater importance to them to have a curriculum which has proven successful, with predictable expenditure of time and effort, than to endure the uncertainties of a possibly more up-to-date schooling which has no precedent. The entrance on the scene of mature engineers, however, who are temporarily freed from the pressures of practice and begin to reflect on the true needs of practice, will provide criticism, incentives and new ideas to the teacher. Hopefully, this example will encourage all students – and in particular the younger ones – to behave likewise and abandon the traditional pattern according to which only the teacher speaks.

## DIRECT BENEFITS OF CONTINUING EDUCATION OF ENGINEERS FOR SOCIETY

In the opening paragraphs of this discussion, a description was given of how the continuing education of engineers serves society in an indirect way by making the training of a vital professional group of society more effective and therefore the results of its work more useful. There are, however, also direct advantages for society at large in the development of such educational programmes for engineers, chief among which might be the fact that the broad dissemination of technical information, which the consumer needs and more in today's technocratic way of life, could be facilitated by the establishment of additional centres of engineering education.

Examples where the uninformed consumer is asked to decide between two alternatives, one of which has short-term advantages and long-term dangers, while the other promises meaningful returns only after longer time-periods are frequent. The outcome of the decision leads generally to the acceptance of the project that in the long run may be the more harmful. In a similar way, the consuming public exerts from the very beginning a direct or indirect pressure upon planning engineers to favour developments and propose alternatives that show only short-term benefits. This means that either the problem must be considered again very soon, or that an irreversible decision has been made whose consequences become more and more undesirable as time goes on. Examples of such decision-making may be found everywhere, be it in the development of traffic or in the area of recreation. The mistakes made demonstrate clearly that the consumer does not understand all the possibilities of participation in the decision-making process. They show furthermore that an uninformed consumer cannot formulate demands and voice his opinion in a responsible manner.

It is well known that the number of decisions which are based on technical facts or triggered by technical developments increases from day to day. If this so-called technological progress is to be truly meaningful to man, each of these decisions must be made as carefully as possible and on the basis of as much significant (technical and non-technical) information as possible. The control of progress depends for its success on the necessary information being gathered and brought to bear in the decision-making.

Much has been said about the process of decision-making, especially where scientific and technological problems are concerned [11]. In this context it is important to emphasize the participation of different segments of society in different phases of the process and to identify these points where the public can influence the course of the decision-making. If one accepts as working hypotheses the differentiation of the following phases of information gathering, information compaction and abstraction (possibly in the form of models), the proposal of alternatives to the question asked, the evaluation of the advantages and disadvantages of each alternative, and finally the selection of one alternative on the basis of quantifiable and non-quantifiable information, then there are at least two phases in which society at large can make its influence felt: first, during the step of information gathering, especially in supplying 'non-technical' aspects of the question at hand, and second in making sure that all available information is considered when the decision-maker selects one optimal alternative. It follows that the non-specialist and outsider must try to obtain as much understanding of a technical project as possible; this point may be illustrated by the discussions about the benefits of 'supersonic transport' in the United States. There an attempt was made to influence decisions about prolonged government support or cessation of the project.

Experience gained in other areas that indicates that the establishment of the increased teaching potential is necessary for the continuing education of engineers will cause an over-all intensification and enlargement of information and teaching services. If one follows the development of centres for adult education or the growing availability of extension courses, one often finds that existing institutes of learning, such as professional schools and universities, have given incentive and support to such projects. They furnish both a stimulating atmosphere and prospective volunteers to satisfy the newly awakened interest and demand.

This interaction may be seen from three viewpoints. First, the technical knowledge that is readily available may increase. Second, there will be a larger number of potential teachers for 'laymen' interested in engineering. Third, the example given by raising interest in general education in engineering topics and then satisfying these needs might demonstrate the necessity of much wider general education facilities and also the possibility of answering such demands. To some it may be a generalization, or sound too idealistic, to say that the establishment of continuing education for engineers will not only pave the way for continuing education of other professionals, but may trigger the development of a much needed system for effective and objective information of society in general. Yet the step from informal organizations grown out of given opportunities, be it the enthusiasm and personal involvement of teachers, or their mere availability, to formal set-ups is not a difficult one and has been made many times. The expectation that society will benefit directly from continuing engineering education through increased dissemination of needed technical information is a realistic one.

While such a flow of information will undoubtedly contribute to a better control of the process of decision-making, the increased availability of information may also lead to a growing-up or maturing of the consumer with respect to his wishes, demands and needs and their satisfaction. New developments in a technical area are frequently started or at least strongly influenced by the wishes or needs of consumers. But, at present, although industry or its R & D department makes an effort to encourage worth-while innovations and to counsel against disadvantageous ones, very often the unreasonable wishes of the public dominate the development.

Yet the efforts so far made to educate the consumer are insignificant. It is here that the increased dissemination of technical information, for instance about the discernible effects of developments that can be encouraged or suppressed, would also show beneficial consequences in the effort to control technical progress. Thus, even before the process of decision-making begins by posing a question, the public should have the possibility or even the duty to decide what questions are to be asked or what needs are the most urgent ones. This must be done by considering short-and long-term advantages and disadvantages even of technical possibilities, not to speak of technical accomplishments. Admittedly, it is still somewhat idealistic to expect from a consumer, who has been told about all the known and projected consequences of a possible technical innovation, that he will act reasonably and deny himself the 'excitement of a new toy'. But this way of controlling technical progress at its very origin seems the most effective and should therefore be chosen.

It has been postulated that the setting-up of continuing education for engineers will lead to increased activity in the area of general education for society at large. Thus the project under discussion will not only serve society indirectly by strenghtening one of its vital professions but may also produce direct advantages in educating the public as equal partners in today's technocratic life. If the various demands of society, with regard to the continuing education of engineers are to be met in the manner described above, society itself must support the project and ensure its realization. For it is eventually the public which will authorize the expenditures, either directly through its governmental representatives or indirectly through private enterprise and other agencies. In this context it must be mentioned that some of the worst failures in American graduate education were due to an inability to pursue steady, consistent policies over the long term. If society wishes to benefit from expenditures in the field of continuing education it must make long-term commitments.

It is hoped that the foregoing considerations and consequent action will also lead to an upgrading of the engineer's image and professional prestige. This would be beneficial for his work and role in society; furthermore an improved image of the profession is needed in order to draw good students into the engineering field which at times suffers from mediocrity. However, it is not only the level of education that some few members of a society have attained which decides the relative inferiority or superiority of one society as compared with another, but the extent and quality of schooling that all the members of a society have obtained. To this end, the establishment of continuing education of engineers can contribute much incentive, experience and practical guidance.

## CONCLUDING REMARKS

It is always a big step from the final agreement to a project and the realization of the idea. The step is difficult because too many interested specialists and public officials have to be consulted about the practical form, the organization, and the administration of such projects. Lengthy deliberations and subsequent negotiations frequently delay action upon a decision that everybody wants to put to work immediately. Thus it is not possible to say anything conclusive about acceptable and workable solutions for the final set-up. There are, however, some possibly useful hints derived from experience at technical university, which may be outlined briefly here. For reasons discussed elsewhere [12], the realization of continuing education of engineers in problem-oreinted research centres that are loosely connected with universities seems promising. Examples of such centres for research and further engineering training exist.[1] The teaching staff preferably consists of a small number of appointed members and a larger number of invited lecturers from practice and from the universities. One of the most important aspects is the development and use of adequate teaching aids, and foremost among them are good publications written for this special purpose. These last considerations are of great importance, since the best concept for an educational enterprise will fail if, for lack of attractive power, the prospective students stay away.

1. For example, Von Karman Institue of Fluid Dynamics, Brussels (Belgium).

*References*

1. V. Berkner, in N. Calder (ed.), *The World in 1984,* Vol. II, Harmondsworth, Middx., Penguin Books, 1965.
2. For instance: '*Umweltbericht' der deutschen Bundesregierung,* 1971 (report on the state of the environment issued by the Government of the Federal Republic of Germany).
3. Hahn, 'Umwelt von morgen', *Umschau,* Vol. 16, 1970, p. 495–500
4. *Information for a Changing Society,* Paris, Organization for Economic Co-operation and Development, 1971.
5. H. Hahn, 'Der Ingenieur von morgen', *VDI–Zeitschrift,* 113/4, 1971, p. 233-7.
6. E. Naudascher, Neue Lehr– und Forschungsziele in der Hydraulik, *Die Wasserwirtschaft,* 61/7, 1971, p. 1-7.
7. H. Hahn, 'Überlegungen zur Berücksichtigung der gesellschaftlichen Auswirkungen der Technik im Studium der Technikwissenschaften', *Studium der Technik-Ingenieure von morgen,* VDI Verlag GmbH, 1971.
8. A. Toffer, *Future Shock,* New York, N.Y., Random House, 1970.
9. H. Hahn, 'Planung der angewandten Forschung auf dem Gebiet der Wasserwirtschaft', *Proceedings of Mannheim Meeting of 'Varantwartung in der Wissenschaft.'* Oldenbourg, München, 1973.
10. J. Fischer, 'Survival University', *Environmental Handbook,* New York, N.Y., Ballantine Books, 1970, p. 134-47.
11. D. K. Price, *The Scientific Estate,* London, Oxford University Press, 1965; or C. E. Lindblom, *The Policy-Making Process,* Englewood Cliffs, N.J., Prentice-Hall, 1968; or H. Hahn, ' "Umweltplanung" – Gemeinsame Aufgabe von Wissenschaftler, Ingenieur und Politiker', *Umwelt,* Vol. 1, 1971, p. 17-23.
12. H. Hahn, "Umstritten: Der Umweltingenieur" Umweltplanung beginnt bei der Erziehung, *Umwelt,* Vol. 1, 1971, p. 36-41.

# Les exigences de la société en matière d'éducation permanente des ingénieurs (résumé)

Professeur Hermann Hahn

L'auteur de cette étude très complète commence par examiner la question du coût élevé de l'éducation en général, et de la formation des ingénieurs en particulier. Pour apprécier la valeur de l'éducation permanente, il faut tenir compte de la nécessisité de conserver à la société la valeur de cet investissement considérable que constitue la formation initiale. Parmi les principaux facteurs à considérer figurent la durée et le contenu de l'éducation initiale, la mesure dans laquelle l'éducation permanente favorise la coopération entre l'enseignement et l'industrie, la diffusion de l'information, et l'influence des progrès scientifiques et technologiques sur l'exercice du métier d'ingénieur. Les préoccupations publiques actuelles concernant l'environnement contribuent largement à activer l'évolution des idées traditionnelles sur la formation des ingénieurs et leur rôle dans la société. La conclusion est que la société va tendre à insister pour que soient créés ou développés des moyens d'éducation permanente et pour que ces moyens soient utilisés effectivement.

# Discussion

A speaker observed that, in Canada, it had been shown that, at least for basic engineering education, the return on the investment was satisfactory. Similar studies had been carried out in Austria.

Several speakers commented on the suggested need for shortening the period of basic engineering education, and it was stated that this could occur if curricula were carefully designed and if there was assurance that continuing education would be available. Whilst such shortening could conceivably prevent students from losing their initial enthusiasm, it should not result from pressures created by student laziness. One possible model of a shortened programme would be to send the student out to practice after an initial programme of mathematics, physical sciences and engineering sciences, and then after, say, two or three years, bring him back to do a programme of social sciences, applied engineering subjects and those aspects of his earlier studies that had become outdated.

Another speaker maintained that continuing education could solve the critical problem of obsolescence of educators, which otherwise can destroy the whole system of education of engineers. The dialogue and contacts created by continuing programmes can solve this problem by forcing teachers to keep up to date and by facilitating this process.

# Content and programmes of continuing education

M. Jerez Juan
V. Obach
I. Pujol Bayó

## GENERAL PANORAMA AND PURPOSE OF THE PAPER

According to various communications presented to the FEANI seminar on the role of universities in post-graduate education of engineers (Prague, September 1970), the status of the continuing education of higher technicians in the principal countries of Europe and America, at the present time, may be summarized as follows:

1. Ordinary engineering courses leading to a first degree, *licence* or ordinary graduation last from three to six years.
2. Higher qualifications such as Master (or more or less similar denominations, e.g. *Diplom-Ingenieur* in Germany, Candidate of Sciences in Czechoslovakia, etc.) and Doctor require additional study courses, which usually last one year at least for a mastership, and two years at least in addition to the dissertation or thesis for a doctorate.[1]

   These complementary studies are followed both by students who have just graduated, and by engineers who have already worked in industry for some years and who thus possess a certain experience. In some countries, special arrangements are made so as to permit engineers to attend the courses without too great an interruption of their professional careers.
3. In traditional university studies, the doctorate constitutes a special preparation for scientific research and professorial activities. Therefore a strict intellectual discipline is required, not only to increase knowledge but also to develop the imagination. Such are the characteristics inherent to the so-called courses of the third cycle.

   While the characteristics just referred to may have some validity in technical studies – although these are essentially directed towards application – the courses leading to technical Master's and Doctor's degrees almost always have a clear aim of specialization and, for that reason, constitute a valid post-graduate activity both for the teaching centre and for the engineers who follows the courses. Nevertheless, these courses are not sufficient by themselves to resolve the problem of continuing education, since they cover

1. In countries where courses for the ordinary degree last five or six years, starting from the level required for university entrance, it is usually considered that diplomas which grant direct access to doctorate studies are substantially equivalent to a degree in enginneering of three or four years plus a Master course.

only a brief period, almost always initial, in the life of the engineer, and finish once the intended degree has been obtained; hence they do not represent a permanent effort that would make it possible for the engineer to incorporate new concepts and methods which have appeared since the completion of his initial studies.

4. Apart from the courses already mentioned, which, by their very nature, belong mostly to the academic field, it is considered necessary to provide courses having as their object the updating of the engineer's knowledge (renovation or recycling courses), thus counteracting the unavoidable obsolescence, and also increasing, improving and diversifying the individual's knowledge in accordance with his interests and professional needs. In the authors' opinion, this kind of course brings the permanent education of the engineer into focus in the most direct and certain way.

The discussion hereunder deals exclusively with the courses referred to in 4 above; we shall call them perfectioning courses for engineers, adopting the term preferred in the United States[1] which is also the more accepted in Spain.[2]

## NEED FOR PERFECTIONING COURSES: IDEAL SCHEME

Although much has been written on this theme, it is worth summarizing here the main reasons for perfectioning courses: (a) they allow the engineer to acquire scientific and technical knowledge which has appeared or has developed since the conclusion of his studies; (b) at different times during his professional career, the engineer may need to investigate the possibilities of techniques with which he is not familiar–questions concerning metallography, semiconductors or managerial techniques would be good examples; (c) at other times, he may find it necessary to transfer from one technical field to another.

Moreover, the education of the engineer on leaving the faculty or college where he obtained his ordinary degree, must be, for obvious reasons, fundamentally incomplete. In fact, it is almost impossible to overcome the difficulty of fitting into the ordinary graduation time the very programmes required today in mathematics, fundamental sciences, basic technologies, the concept of design and its techniques, organization, economics, etc. If these subjects are to be gone into deeply, then specialization must inevitably be restricted. Consequently it seems unavoidable that the ordinary degree phase will prepare the engineer to operate only at the minimum level required, and that he must rely for further education, information and training upon higher degrees and perfectioning courses.

It seems appropriate to reproduce here a statement which appears in the communication to the Prague seminar by the Director of the National Agronomic Institute of Paris and President of the Commission for Engineering Titles in France, M. F. Roger Blais:[3]

1. See the report presented by Professor Joseph M. Pettit, Dean of the School of Engineering, Stanford University, California, and elect President of the American Society for Engineering Education, at the third general congress of WFEO/FMOI in Varna (Bulgaria), June 1971.
2. The centre, established in Barcelona (Spain) on a permanent basis, organized and maintained by the National Association of Industrial Engineers, is called the Centre for Perfectioning of the Engineerr (CPI).
3. See the proceedings of the FEANI education seminar,1970, held in Prague.

Pedagogy is evolving its techniques, syllabuses their contents. The College should more form than inform and have confidence in permanent education, so that the pupil-engineer of today may be up to date in the year 2000.

On the basis if this paragraph, we think that the ideal aim of permanent education is to ensure a sustained development of the aptitudes of all engineers throughout all their professional lives. For many years it has been accepted as a general rule of life that every man must study until a certain age; work until another age, more or less determined, and then rest. It seems possible that the interest that society shows nowadays in continuing education will somewhat change that concept, and will tend to bring about big changes in the time distribution. We believe that prospects for the immediate future will be marked mainly by the following development: the admission of recent graduates or even students into industrial organizations at an age lower than at present; a greater dedication of individuals to educational activities throughout their professional lives, and, at a certain age, a gradual enlargement of leisure time. In this way it seems possible to prevent the two great changes which every professional man or woman has to undergo (and which in many cases cause a psychic trauma), namely passing from university to work, and from activity to retirement.

## AN ATTEMPT TO CLASSIFY PERFECTIONING COURSES FROM VARIOUS POINTS OF VIEW

We present hereunder an attempt to establish a classification of these courses, restricting ourselves to those which do not lead to a higher degree, since other participants have given full and precise information on that type of course at this and the previous seminars.

### *According to the kind of organization which prepares and develops them*

Three kinds of courses may be distinguished: (a) courses given by faculties or colleges of engineering; (b) courses organized by research institutes; and (c) courses which are run by institutions other than those already mentioned.

With reference to the third of the above groups, one must distinguish between the non-profit courses organized by professional organizations or industrial or economic boards or associations and those given with the purpose of obtaining a gain or promoting the use (and, indirectly, the commercial publicity) of certain machines or equipment.

### *According to the professional maturity of the persons who attend*

Certain courses are addressed almost exclusively to recently graduated engineers with the aim of completing their education or to initiate them into a speciality, whereas others are designed for experienced practitioners.

### *According to the subject matter*

Courses which are aimed at improving background knowledge may be distinguished from those which are directed mainly at increasing the student's knowledge and experience of a particular branch or speciality.

### *According to the length*

Both long and short (or abbreviated) courses are offered. It is neither possible nor worth while to fix precise limits for these two categories. As a rule of thumb, courses lasting six months or more may be considered long, and short courses may be understood as those with a duration not exceeding four weeks. An intermediate category could also be considered.

### *According to the premises used for the teaching*

These include (a) a traditional classroom; (b) a room laid out rather like a council meeting room so as to situate the students at the same level as the teacher and thereby improve the tendency to participation; (c) classroom and laboratory; and (d) classroom and insutry.

In this connexion, remarkable experimental course of the university-industry type with a long duration and having substantial means at its disposal may be mentioned: this is to Bosworth Course, organized in recent years in the United Kingdom and which seems to have been highly successful.

### *According to the means used*

These include teaching with blackboard and chalk; the same but complementing lectures and discussions by audio-visual aids or demonstations in the same room; the use of remote facilities by means of a micro-wave link with one-way video signals and two-way audio communications. Some of the above will be commented on in the following paragraphs.

## CONTENTS AND PROGRAMMES OF THE COURSES

In the preceding section, we distinguished between courses aimed at improving the engineer's basic knowledge and those which are intended to give a thorough knowledge of a particular speciality. Here we shall examine in some detail the contents of typical courses of both types.

### *Information courses*

These information courses are similar to those at the university. The method of teaching is chiefly unidirectional, e.g. the teacher is at a higher academic level than the pupils. The teacher communicates his knowledge to the pupils who learn a subject of which they had little or no previous knowledge.

This type of course may be appropriate in the following cases: (a) young engineers who wish to acquire knowledge of subjects not taught in the faculty or college; (b) engineers who wish to acquire complementary techniques with a view to promotion; (c) engineers who are obliged to change their speciality for various reasons.

### *Perfectioning courses with a two-way flow of information*

Here, the levels of knowledge of teacher and students are comparable, and the flow of information alternates from one to the other.

In perfectioning courses time is a fundamental variable. Anything that can be done outside the classroom should not be done within it. The time devoted to lectures in perfectioning courses tends to dimininsh, amounting usually to between twelve and thirty hours. At the end of a course, the general opinion of the engineers attending it is that it should have lasted longer; when asked, however, whether they would have been able to attend if it had been longer, many of them answer negatively. The teaching methodology in a two-way type of course requires a great effort on the part of the teachers. To the careful preparation of the lectures, which are needed in courses of this kind also, must be added the permanent effort necessary to encourage participation while, on the other hand, they must ensure that the time-table for the programme is respected.

Here we meet again the question already mentioned and considered as fundamental regarding the methodology suitable to these courses, namely that of time. It is well known to people concerned with educational activities that, although active methods possess real advantages, they present the difficult problem of fitting oral expositions, discussions, etc., on whatever subject, into strict time limits

In courses of this kind, even if they are of the recommended length of from twelve to thirty hours, it is better to have more than one teacher in charge. In our opinion, the ideal number is two or three. One teacher may fatigue the listeners, and also the teaching may be greatly influenced by his personal point of view. On the other hand, more than three teachers in the same course could convert it into a series of unconnected lectures with the dangers of overlapping and lack of illation. An average of eight to ten hours per teacher seems to represent the optimum.

It is not considered advantageous for perfectioning centres of a non-university character to maintain a permanent body of teachers, but rather to select them for each subject among specialists of recognized authority and aptitude. In this way, the teachers are given a professional challenge which, in our experience, is normally accepted. The economic reward does not usually present an obstacle, as the amount that can be offered for a lecture time of eight to ten hours represents only a small percentage of the teacher's annual income; outstanding professionals do not attach too much importance to this point as a rule.

*Subjects*

The recommended length of from twelve to thirty hours limits the selection of subjects to those that can be covered thoroughly, in such a short time. It follows that the teaching of basic subjects is automatically excluded.

For example, it would make little sense to offer a perfectioning course with a title like 'Electricity'. On the other hand, titles such as the following seem to be commendable: 'Earthing of Electric Plant', 'Lubrication in the Electrical Industry', 'Economic Optimation of Transformers', 'Transitory Phenomena in Electric Machines', to mention only four.

Our experience shows that the more restricted the subject, the greater is the interest shown by the engineers concerned. In the perfectioning courses given at CPI, the greatest successes were obtained with subjects of small amplitude, allowing a thorough development on the part of the teachers and a detailed understanding on that of the students.

*Time-tables*

The information we have received concerning different countries shows that there are great variations in time-tables. Some examples are: (a) full-time courses lasting several weeks or several days; (b) part-time courses lasting from fifteen to forty hours; (c) courses of one afternoon a week for several weeks; (d) courses of one complete day per week for several weeks; (e) courses of five complete consecutive afternoons; (f) week-end courses, i.e. on Friday, Saturday and Sunday; and (g) courses given in the early hours of the morning. Time-tables depend very much on the situation of the teaching centres and on the support granted by the management of the enterprises concerned. Centres located in small towns or in isolated places are obliged to convene full-time courses. However, centres in large cities or industrial concentrations may organize their course with any convenient time-table.

From the characteristic profile of the student given above, some conditions may be inferred with regard to the staff and the pedagogical methods which are advisable. Although we shall not go deeply into this field, since it belongs properly to another section of the seminar, we must devote some words to it.

The master must be a real expert in the theme, with completely up-to-date knowledge, considerable experience in the field to which the subject refers and, if not a professional teacher, he must have the capacity of communicating his own learning and experience to others.

The methods applicable to this kind of teaching must be based on the criteria of conciseness and efficiency, and must take fully into account the special features of the student engineers, whose previous education and quick comprehension of scientific reasoning may be assumed. We consider it of great importance that, at least twenty-four hours before each lecture, a clear and concise summary of the proposed exposition, in schematic form if possible, should be distributed among the students, with the aim of directing their attention towards its contents and facilitating their thinking on the points which deserve to be enlarged or discussed.

We recommend that all explanations and discussions be conducted in such a way that the need for note-taking is eliminated or reduced to a minimum, since this secondary activity tends to disperse the attention of the listeners.

With respect to techniques of programmed learning, nowadays so widely used, we think that they represent a useful and adequate device when a teacher is confronted with one single student, or as the basis of the so-called auto-learning. However, these methods present considerable difficulties for groups of listeners, however small, and for this reason their adoption is not recommended in perfectioning courses for engineers. Finally, we think it worth while to mention the technique which consists in recording a course with images and sound on a magnetic tape.[1]

We have been informed of the very interesting programmes of certain courses given in the United Kingdom and in the United States on a full-time basis during a whole year, especially dedicated to engineers who finished their studies at least ten years before.[2] Although this system is theoretically sound, its practical application presents many difficulties. First of all, only a small number of firms

1. See J. M. Biedenbach, *Applications of Video-bands to Programmes of Perfectioning Studies for Engineers* (exposition to the seventy-seventh meeting of the American Society for Engineering Education, 1969).
2. It is possible that courses of the same kind have been organized in other countries also.

are willing to sacrifice the services of an engineer during a whole year, paying his full salary, meeting the cost of the course and accommodation, and making sure that he is not disturbed by the problems which arise at the place of work. The institution organizing the course could invite the companies which intend to enrol one more of their technicians to sign an undertaking to fulfil the above conditions, as has been done on some occasions by MIT in the United States. However, this could produce a supplementary difficulty which should be taken into account.

In our opinion, the above comments justify the trend towards abbreviated courses as being the most practical and effective in perfectioning education, considering all the circumstances which concern the engineer and the medium in which his professional career develops.

## EDUCATIONAL METHODS

The typical profile of the student attending a perfectioning course is that of an engineer with the following characteristics: (a) he is engaged in a professional activity related to the subject of the course; (b) he wishes to improve his knowledge of that theme and bring himself up to date; (c) he is willing to discuss and compare his knowledge with that of his colleagues working in the same speciality; (d) it is quite feasible that, at certain points of the programme, his knowledge and experience will surpass those of the teacher; (e) he may propose a particular problem or question for study or solution during the course. For instance, the Colorado University in the United States has already recorded some courses in this way. The reproduction of such recordings before groups of engineers with the help of a teacher familiar with the recorded subject and prepared to clear up any questions or doubts that may arise, is considered as highly efficient. The present high cost of video-bands will, of course, limit the use of this technique; however, the possibility of repeating the course for different groups may compensate the initial expense.

## RECEPTION OF PERFECTIONING COURSES ON THE PART OF THE ENGINEERS

Although certain countries have made it compulsory for engineers to participate in perfectioning or similar courses, it must be recognized that, as a general rule, participation is not very high. We are still a long way from the situation referred to on page 102 as an ideal scheme.

The authors believe that organizations like Unesco and FEANI ought to promote a large-scale campaign to encourage the engineers of all countries to devote at least 1 per cent of their professional time every year to follow courses and seminars.

Some of the bodies which promote courses of continuing education in the United States inform us that they consider it a good attendance if 7 or 8 per cent of the engineers in an influential region or zone attend a course or seminar in the course of a year. In Spain, in the academic year 1970/71, CPI (Barcelona) succeeded in obtaining an attendance of 5 per cent of all the engineers residing in Catalonia.[1]

At present, it seems that interest in continuing education is increasing rapidly all over the world, which gives grounds for hoping that good results will be obtained during the next decade. The suggestion that engineers should devote 1 per cent of their professional time annually to continuing education could be put into practice by attendance at a short course each year or a longer one every two or three years.

## SOME CONCRETE PROGRAMMES

The following is a list of the titles of some courses given recently at CPI.[2]

Introduction to Electronic Computers
Purchasing Policy
Analysis and Planning of Maintenance Work
Automation in Maintenance
Introduction to Modern Mathematics
Industrial Painting
Water Treatment for Steam-boilers
Industrial Costs and Budget Control
Application of Electronic Computers to Five Cases
Treatment of Sewage
Maintenance of Industrial Plants
FORTRAN: An Instrument for Calculation
Banking and Credit Operations
Ergonomics
Numerical Control; Background and Applications
Furnaces, Burners and Fuels
Economic Optimization of Electric Transformers
Production and Utilization of Steam
Thermic Treatments
Psychology for Directors and Leaders
Foundations and Earthworks
Knowledge of Paints
The New Spanish Standard for Concrete Works
Matri-calculation of Metallic Structures
Urban Traffic
Reliability
Integrated Linear Circuits
Solid Handling
Transitory Phenomena in Electric Machines
Current Techniques for Spinning Cotton and Mixed Fibres
Thermofluids and Super-heated Water
Cooling in the Food Industry
Electric Welding by Resistance Methods
Refractory Products and their Applications
Lubrication in the Electrical Industry
Industrial Use of Gaseous Fuels
Physical Distribution of Goods
Pumping Equipment

# Contenu et programmes de l'éducation permanente (résumé)

M. Jerez Juan, V. Obach, I. Pujol Bayó

Les auteurs traitent des matières que doivent comprendre les cours donnés dans le cadre de l'éducation permanente et partant de l'idée que, lorsqu'un ingénieur

1. Region in the north-east of Spain, with a comparatively high industrial development.
2. The majority of these courses were completed in twenty hours. The attendance was limited to twenty engineers.

reçoit son diplôme, sa formation est fondamentalement incomplète. C'est pourquoi il lui faudra suivre pendant toute sa carrière divers cours de "perfectionnement" pour passer d'un domaine technique à un autre. Les cours diffèrent selon le genre d'établissement où ils sont donnés, le niveau de ceux qui les suivent, le sujet, l'abondance des matières à assimiler et donc le temps nécessaire, ainsi que selon les moyennes d'enseignement utilisés. La conclusion est qu'un effort de grande envergure est maintenant nécessaire pour faire en sorte que les ingénieurs participent davantage aux programmes d'éducation permanente, et des organisations comme l'Unesco et la FEANI devraient aider à promouvoir cette participation accrue.

Le document se termine par une liste des cours récemment donnés au centre où enseignent les auteurs.

# Discussion

Several speakers brought up in the discussion the question of where continuing education should be carried out, in the universities, in industry, in special centres, or by engineering societies.

With regard to content of programmes, the studies being carried out in the Federal Republic of Germany by VDI were described, notably the work on defining qualification profiles and job profiles. The programmes of the VDEh were also described. The work in continuing education being done in France by such bodies as the Society of Electrical and Electronic Engineers (SEE) was mentioned.

One speaker suggested that longer courses, up to several months in duration, are suitable for younger engineers and for teaching management, in depth, but that they would usually be of short duration for more senior personnel. He also brought up the question of the methods used for choosing topics for courses, and in reply Dr Obach said that in his centre a panel of engineers had this responsibility, and that questionnaires to participants were also used. This was seen by several speakers as being part of the general question of the location of responsibility for taking initiatives in organizing continuation education, and of feedback from the recipients within the profession. There seemed to be general agreement that engineering societies have a key role to play in this matter.

The amount of time that should be spent by an engineer on his own continuing education was mentioned and the figure of 1 per cent advanced by Dr Obach was compared with that of 15 per cent mentioned earlier. The latter figure was seen as comprising all types of study, including reading technical publications inside and outside normal working hours, seminars, and other self-study in fields such as economics or the social sciences. The actual field of work of the engineer was seen as being an important factor, as an engineer engaged on research and development would probably have to spend more time on continuing education than an engineer engaged in, say, production and maintenance type of work in industry or in civil engineering construction.

# Planning of training programmes using job analytical methods

Kurt E. G. Gestrelius

By 'job analyses' is meant the process by which the tasks involved in a particular job are clarified, while 'job description' denotes the results of the job analysis.

The term 'target group analysis' describes the process used to investigate the knowledge, skills and attitudes of the presumptive course members at the start of the training. The results of the target group analysis are called 'target group qualifications'. The job training should naturally cover only the knowledge, skills and attitudes that prospective employees lack prior to the training.

The type of analysis of training needs dealt with here is based on a very simple model, in which people are placed in situations the mastery of which can sometimes require training. These situations can arise during working hours or in leisure time. The model has been used as a basis for the majority of psychological theories. It makes use of three different kinds of variables, namely stimulus variables, inner process variables and reaction variables (see Fig.1)

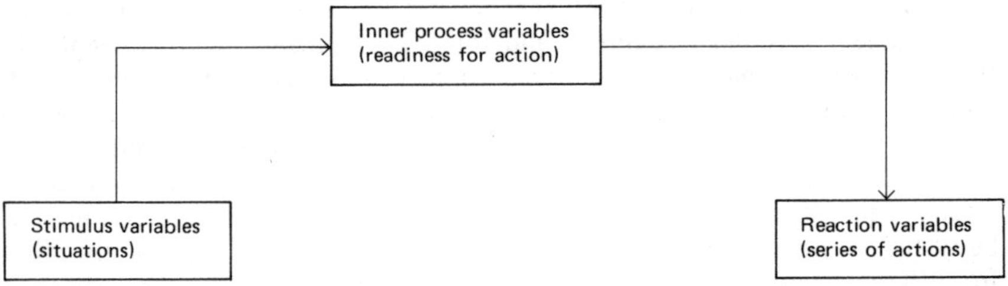


Fig. 1.

According to this model, the individual is imagined in a series of 'problem situations' and his actions bear witness to the way in which he tackles the problems.

Job analyses can establish what tasks the employees are faced with, and the tasks can then be transformed into knowledge, skills and attitudes that it is necessary for the job holder to have. Target group analyses can give information about the target group qualifications in question. Comparison of job descriptions and target group qualifications can provide a basis for assessment of the training needs.

## TRAINING PROFITABILITY

Another concept related to training needs is training profitability. Factors that affect training profitability are: (a) the average length of time that employees need to serve in their appointments without training in order to reach the lowest acceptable level of efficiency in the various tasks involved in the job; and (b) possible demands that the employees should, from the very beginning, work at a particular level of efficiency.

Society has some jobs for which extensive training is required and others for which no training at all is required. A scale can be imagined lying between these two extremes, along which the training needs for different jobs can be expressed in terms of the length of training. One should beware, however, of believing that the need for special job training is greater for jobs involving rights and obligations of great importance for society. The majority of executives in the business world and the general administration, for example, have for a long period not received any compulsory job training. The reason for this is naturally that those responsible for recruiting people for jobs of this type try to choose those whose personal qualities and earlier education and experience fit them for the job in question.

This shows that staff recruitment also influences training needs and training profitability. If it were possible to recruit personnel whose previous education and experience had given them the attitudes, knowledge and skills necessary for efficiency in the job, or at least in part of it, the training need would be changed to a corresponding degree. For various reasons, however, the opportunity of recruiting staff from the groups requiring least training is often very limited. But it is essential for deciding training needs and training profitability to establish the attitudes, skills and knowledge of the presumptive employees in relation to the demands made by the jobs.

## JOB ANALYSIS AND TARGET GROUP ANALYSIS – TWO STAGES IN THE SYSTEMATIC PLANNING OF TRAINING

In the discussion that follows, a job is understood as being something separate from the employee. Jobs are thought of as being the social positions which are composed of the obligations and rights bound up with them. Employees are those appointed to take on and discharge these obligations and rights. The way in which the employees act in their jobs represents the dynamic aspect of the jobs. This aspect means that the profile of the job is also considered to be dependent on the personalities of those who take on the obligations and rights of the job.

Obligations and rights can for different reasons be associated with a job. They can be associated with a job by statutes and directives or by practice established over a period of time. But the view taken of a job also depends upon what different employees feel about these rights and obligations. Apart from the employees themselves, other groups in society can also have ideas about the rights and obligations pertaining to a certain job. Thus it is not only by means of statutes and directives, which can often be interpreted in different ways, that society can express what it expects of those who shoulder the rights and obligations of a particular job. All these expectations, including those of the employee himself, will influence not only the rights and obligations that are ascribed to a job but also the ways in which the employee is going to function in the job. This approach is very important when one wants to gather information which can be used for describing a job. The goal of the job analyses is in this context primarly to provide information about the tasks facing the holders of the respective jobs. The job analysis has the further aim of clarifying as far as possible the way in which the obligations and rights of the job tend to develop.

A target group analysis can be said to have two sides: one dealing with the course members capacity for study, which is of significance for the didactic planning of the course; and one dealing with the content of the training, finding out which items the course members already command satisfactorily. In this study, the target group analyses are seen primarily as an aid in deciding which parts of the obligations and rights of the jobs the presumptive employees have sufficient experience of at the start of the job training.

Job analysis and target group analysis can be seen as stages in the extraction of training goals. The general planning process, of which the extraction of planning goals is a part, forms what is known as a training technological pattern, as summarized in Figure 2.

Figure 3 presents the items that can be included in an extraction of training goals. It can be seen how target group analysis and target group description on the one hand, and job analysis and job description on the other, have been used in deciding the training goals.

Job analyses and target group analyses lead in their turn to several other tasks. An example of a process the main features of which can be used for both job analysis and target group analysis is given in the first four parts of Figure 4.

Taking the survey presented in Figure 4 as a starting-point, the tasks involved in planning training for teacher-trainers can be summarized as given below.

### *Deciding types of information*

The question of the type of information to be collected can be tackled from several aspects. In general, of course, it should be kept in mind that the purpose of the job analysis should decide what information is to be collected. Relevant material will then be what can be used as a basis for training goals and training content. But when this point has been agreed upon, there still remain to be answered questions concerning what type of information is of value as a basis for the goals and content of the training. Aspects of this kind are, for example: (a) previous knowledge; (b) difficulty; (c) expectations; (d) time; (e) environment; and (f) the future. As far as previous knowledge is concerned, it is a question of whether one has at the start a good enough idea of what this particular job demands of those holding it to be able to seek information within certain main categories of these demands. If there is thought to be a risk that this kind of controlled data collection might lead to some parts of the job being neglected, one can decide to seek information unconditionally. The difficulty aspect concerns the possibility of focusing the data collection on material that sheds light upon the particular difficulties that the employees and others feel to be associated with the job.

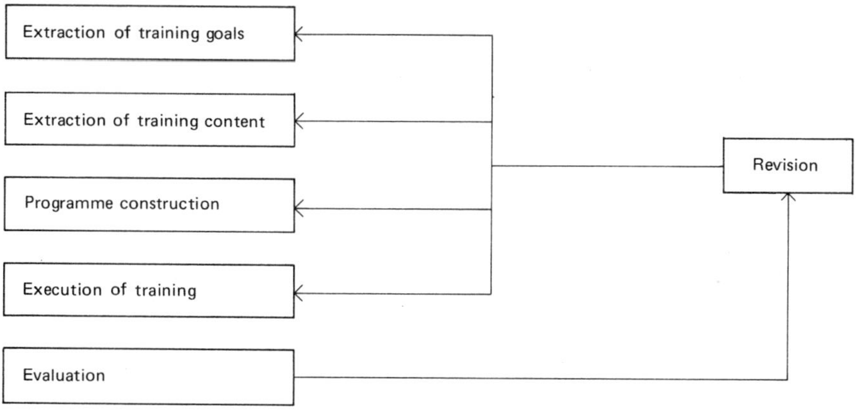


Fig. 2. *General model for the planning, execution, evaluation and revision of job training.*

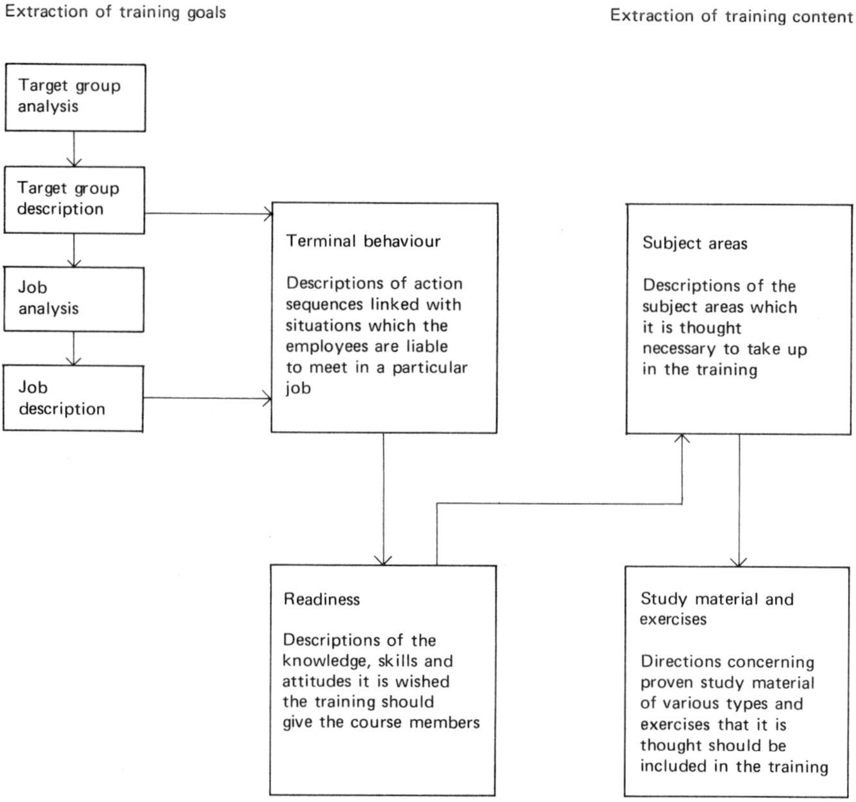


Fig. 3. *General model for extraction of training goals and training content when planning job training.*

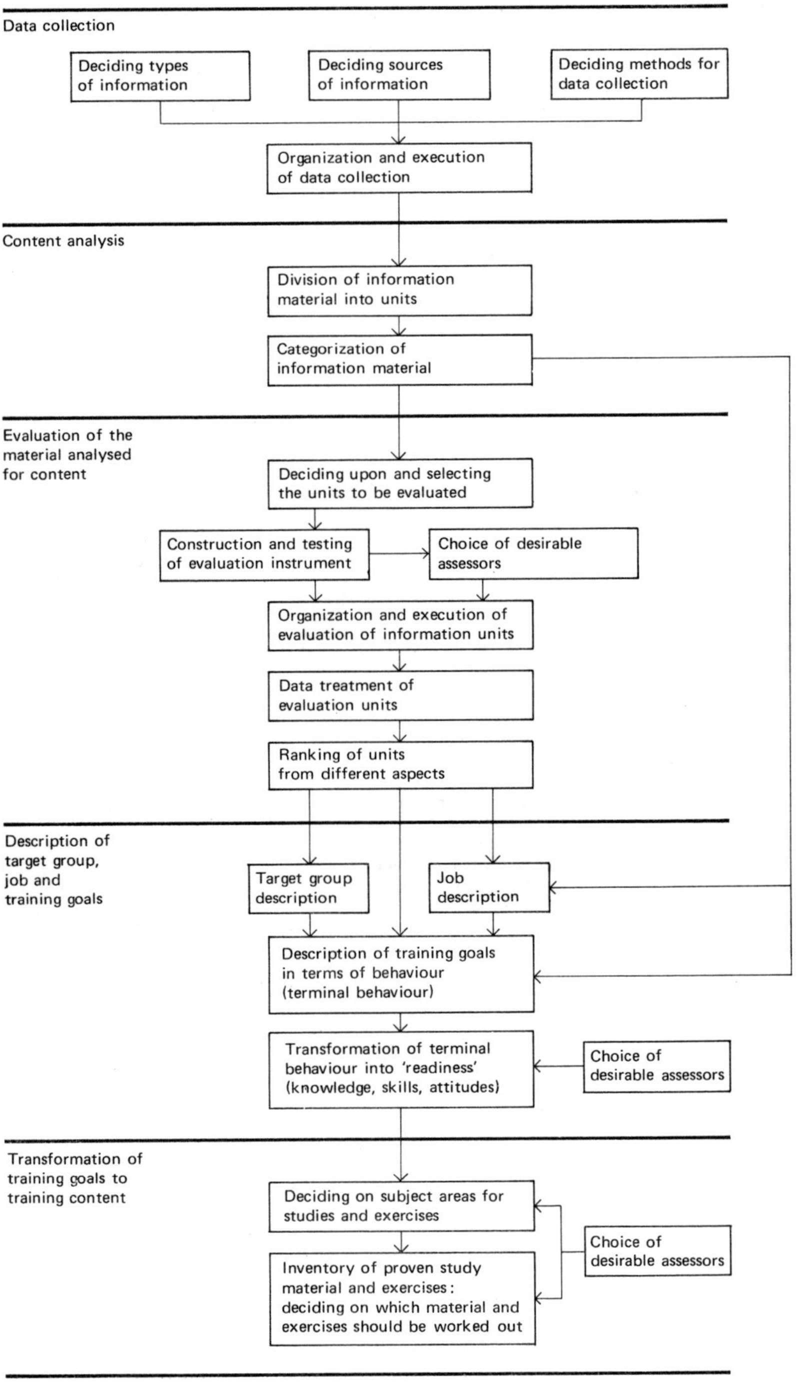


Fig. 4. *Processes for planning job training, up to and including a decision on training content.*

Expectation refers to the difference between expected and actual tasks involved in the job. The employees can, for example, be asked to express what tasks they personally think they should be given instead of the tasks they in fact have.

The time aspect concerns whether when collecting data one should seek information not only about what tasks the employees have, but also about how great a proportion of their working time is occupied by different types of tasks.

Environment can mean requests for information about the employees' (a) personal contacts; (b) means of communication; and (c) place of work. The future aspect involves focusing the data collection on conditions concerning the developmental tendencies of the jobs and probable changes in the foreseeable future.

### *Deciding on sources of information*

When planning the job analyses and target group analyses, it is a great help to find out which sources of information are conceivable for the jobs in question. Useful sources of information can be (a) official and unofficial texts (statutes, directives, etc); (b) expert groups; (c) observers; (d) employees themselves; and (e) the employees' contact groups. The decision as to which source or sources of information to use is usually dictated by the amount of time and money available for the analyses.

### *Deciding on the methods for data collection*

There are two aspects to the question of which methods should be used for collecting the data required for a job analysis. One concerns which methods are most suitable with regard to the type of information one is looking for, the other which methods should be used with regard to the source(s) of information one intends using. The limitations of time and money and access to personnel can naturally also influence the choice of methods. These restrictive factors influence not only the choice of method but also the number of people asked to provide information. The following methods appear to be useful for collecting data for the job analyses: interview, questionnaire, diary, observation, analysis of text content, analysis of 'flow of business' and practical experience. The main advantage of the interview is that the interviewer can stimulate the person being interviewed to give a large amount of detailed information. The problem of registering long interviews can be eliminated with the help of tape recorders. By using a postal questionnaire, it is possible to contact a lot of people at a low cost, but there is usually a certain non-response.

Observation methods can be divided into direct observation and self-observation. By direct observation is meant observation carried out by special observers, while self-observation means that the employees observe themselves and report their tasks, usually in the form of a diary.

Where the observation of intellectual work is concerned, there is often little possibility of collecting information about the actual content of the job by means of direct observation. In such cases, the job content is said to have low observability.

## CONTENT ANALYSIS

Whatever methods have been used for collecting data, and however the acquired information has been registered, it must be analysed and transformed into a definite and manageable form. The analysis of content can be carried out in two stages. The first stage involves dividing the material into certain units, e.g. tasks (situations). The second stage involves (a) construction of a category system and (b) classification of the units of the information material.

An example of a category system for teacher trainers is given in Figure 5.

| | Basis for division content | | | | | | | |
|---|---|---|---|---|---|---|---|---|
| Basis for division 'contact' | Student analysis and evaluation 1 | Goal analysis 2 | Material analysis 3 | Programme construction 4 | Execution of training 5 | Own training 6 | Organization and co-operation 7 | Active school service 8 |
| School authority A | | | | | | | | |
| Colleagues B | | | | | | | | |
| Department of education C | | | | | | | | |
| Tutors D | | | | | | | | |
| Student teachers E | | | | | | | | |
| School students F | | | | | | | | |
| Teaching practice school G | | | | | | | | |
| 0 contact (null contact) H | | | | | | | | |
| Meetings I | | | | | | | | |
| Sundry K | | | | | | | | |

Fig. 5. *Example of a category system for analysis of teacher trainers' jobs.*

This system of categories consists of two different bases for division. The categories in the bases for division must naturally be described exactly. The figures in the squares of the category system state how the information material in question has been classified. Starting from this type of classification, one can give a short presentation of what the information gathered from the different sources of material is about (see Figs. 6 and 7).

## EVALUATION OF THE MATERIAL CATEGORIZED

If job descriptions and the material analysed for content are to be used in planning, carrying out and evaluating training, the work must be divided into

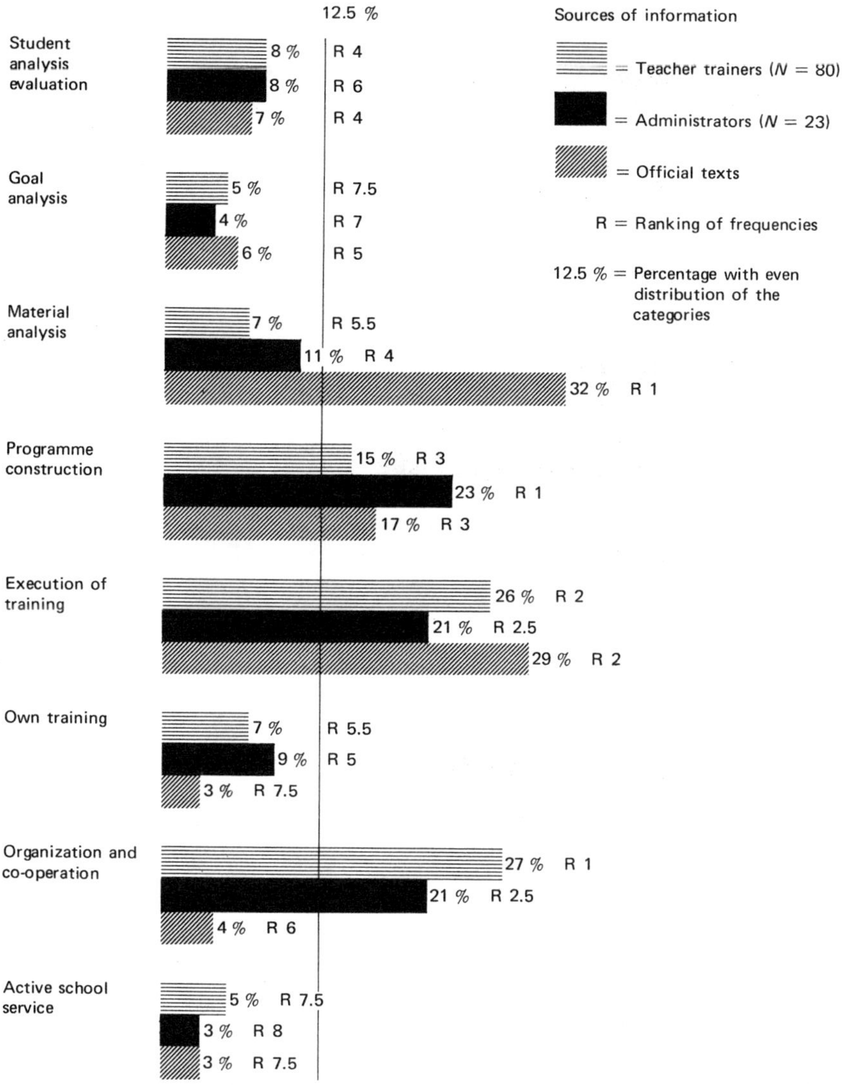


Fig. 6. *Example of presentation summarizing what information material from different sources deals with. In this example the material has been divided into 'content categories'. The material concerns teacher trainers.*

several stages. The way in which those planning the training tackle the tasks involved in each stage will be decisive for how the job descriptions and the material analysed for content can be used in the actual training. Considering that both content analysis and job analysis demand a great deal of time and work, it is surprising that so little has been written so far in educational publications about the working procedures necessary if the training is to make full use of these analyses.

So far no reservations have been made concerning the number of situations that it is possible to make use of when describing the jobs and the training goals.

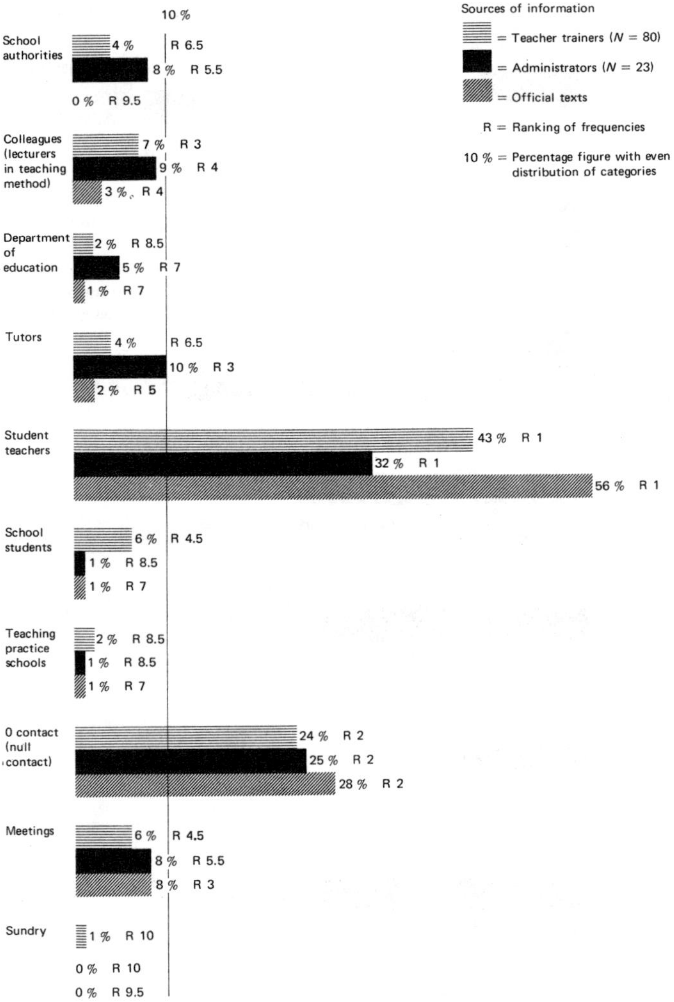


Fig. 7. *Example of presentation summarizing what information from different sources deals with. In this example the material has been divided into 'contact categories'. The percentage figures state to what extent a job's tasks are said to require contact with different personnel groups. The material concerns teacher trainers.*

In order to be able to decide which situations require training, the situation material should preferably be evaluated from certain aspects. If the material contains a very large number of situations, it can be impossible for practical reasons to evaluate all of them. Therefore it is necessary to choose a method for selecting the number of situations that can be assessed with the resources available. One can, for example, use questionnaires as the evaluation instrument. Since the methods for collecting data require so much time and work, the number of people providing information is usually strictly limited. In this way one has the chance of planning an evaluation stage so that it can encompass many more people than just those providing information.

The classified situation descriptions can be evaluated from different aspects. Since in this case the work concerns training it is obvious that priority should be given to the aspects connected with training needs. Examples of evaluation aspects and assessment scales are presented below, together with comments intended to elucidate the import of the aspects for those responsible for the evaluation.

*Aspect A*

This is a situation I have been faced with . . . . . . . . . . . . . . . . . . . . a
This is a situation I have *not* been faced with . . . . . . . . . . . . . . . . . b

*Comment*

In this aspect it is a question of deciding whether or not one really has met with the situation described.

*Aspect B*

This is a situation that I ought to be faced with . . . . . . . . . . . . . . . . . c
This is a situation that I ought *not* to be faced with . . . . . . . . . . . . . . d

*Comment*

In assessing from this aspect, you should assume that the situation can occur, whether or not it is in itself desirable. Imagine that the organization where you work is such that situations that you consider you should not be faced with can be dealt with by others.

*Aspect C*

From the point of view of the *present* objectives of the job, has the situation . . . . . . . . .

No significance 0
Little significance 1
2
3
4
5
6
7
Great significance x

The situation *cannot* be assessed from this aspect.

*Comment*

The aspect should be assessed on the basis of how significant you yourself feel the situation to be in relation to the objectives you are trying to attain in your job.

*Aspect D*

From the point of view of the *future* objectives of the job, the situation has . . . . . . . .

No significance 0
Little significance 1
2
3
4
5
6
7
Great significance x

The situation *cannot* be assessed from this aspect.

*Comment*

The aim with this aspect is to get an idea of how significant different situations will be in relation to the objectives you yourself as a teacher trainer would like to attain around 1980 (bearing in mind the probable development during the 1970s).

*Aspect E*

| | | | |
|---|---|---|---|
| When I am faced with this situation, it usually causes . . . . . . . . . . . . . . . . . | No problems | ☐ | 0 |
| | Minor problem | ☐ | 1 |
| | ↓ | ☐ | 2 |
| | | ☐ | 3 |
| | | ☐ | 4 |
| | | ☐ | 5 |
| | | ☐ | 6 |
| | | ☐ | 7 |
| | Great problems | ☐ | x |

The situation *cannot* be assessed from this aspect.

*Comment*

By problems we mean that considering the demands one makes upon oneself one can sometimes experience to a greater or lesser extent a feeling of inadequacy when faced with a particular situation.

*Aspect F*

| | | | |
|---|---|---|---|
| Considering the total annual work-load involved in the job, this situation demands, *from the point of view of time* . . . . . . . . . . . . . . | No work | ☐ | 0 |
| | Little work | ☐ | 1 |
| | ↓ | ☐ | 2 |
| | | ☐ | 3 |
| | | ☐ | 4 |
| | | ☐ | 5 |
| | | ☐ | 6 |
| | | ☐ | 7 |
| | Much work | ☐ | x |

The situation *cannot* be assessed from this aspect.

*Comment*

By work-load from the point of view of time is meant here a rough estimate of the total consumption of time that the situation demands during a year.

*Aspect G*

| | | | |
|---|---|---|---|
| In this situation, *with the experience I now have*, I need . . . . . . . . . . . . . . . . . . . . | No training | ☐ | 0 |
| | Little training | ☐ | 1 |
| | ↓ | ☐ | 2 |
| | | ☐ | 3 |
| | | ☐ | 4 |
| | | ☐ | 5 |
| | | ☐ | 6 |
| | | ☐ | 7 |
| | Much training | ☐ | x |

The situation *cannot* be assessed from this aspect.

*Comment*
In this aspect you should assess to what extent you would deal with a situation even more smoothly and efficiently if you were given some form of training in mastering it.

It can be seen above how those assessing the situations worked with an eight-point scale for aspect C to G. Aspects A and B are mainly intended to be used to clarify the developmental tendencies of the tasks involved in each job. The aspects chosen can be motivated as follows. Situations getting high values for both aspects A and B can be said to be highly topical at the present time. Situations getting low values on aspect A but high values on B appear to be 'coming situations', while those getting high values on A but low values on B could be called 'receding situations'. Situations that get low values on both these aspects appear to be of little interest in connexion with the planning of training.

Aspects C and D can be looked upon as control aspects of the reasoning above. Thus in principle the C aspect should provide information about the significance of situations that get high A and B values or alternatively high A but low B values. The D aspect is correspondingly intended to give information about the future significance of situations with high A and B values or alternatively low A but high B values. The E aspect can be said to be a control aspect of the G aspect. Thus there has been a general expectation that situations with high E values should also get high G values, but situations with considerable differences between the E and G values are naturally also conceivable and such situations are also of great interest for the planning of the training. The F aspect should be looked upon mainly as an aspect that can provide supplementary information about situations that otherwise demonstrate similar training needs. The aspect can also, however, give information that will be useful in backing up descriptions of the present content of the jobs. The F values can be combined with the distribution of the total material over the content categories and can thus provide a better basis for descriptions of the characteristic features of the jobs.

Each situation can be given a ranking for each aspect. The different rankings of the situations can be combined in different ways and on this basis new rankings can be obtained. Examples of combinations of the rankings of the aspects that can be used for extracting the situations of the greatest interest for the training are the following.

1. A + B + C + D + E + F + G; the combination can be called 'all'.
2. B + D + E + F + G – A – B; the combination can be called 'coming training needs'.
3. E + G – A – B; this combination can be called 'unusual problems requiring training'.
4. E – G; this combination can be called 'problems not requiring training'.

In the combination 'all', the seven aspects have all been thought equally important, the ranking of each situation can be added and from these ranking totals new rankings can be calculated.

In extracting the situations of the greatest interest for the training, one is naturally very interested in situations that have got assessment values that are of particular interest for the future. The combination 'coming training needs' is composed in such a way that high rankings are dependent on B, a large number of assessors considering that they should be faced with the situation, relatively high values for D the future importance of the situation, E the problems it brings with it, F the work required in terms of time, G the training needs it involves and on A, few assessors having claimed to have met the situation, and low values for B the present significance of the situation.

Situations getting high rankings for the combination 'unusual problems requiring training' appear to be of particular interest for the training and these situations can be used to supplment those extracted earlier in so far as they have not already been in accordance with the principles given above.

Situations getting high rankings in the combination 'problems not requiring training' are of special interest for anyone planning training since they show which problematic situations the assessors do not think can be made easier with the help of training. Those assessing the situations perhaps believe that it is impossible to be trained to deal with them, since they themselves have never received such training.

## TRANSFORMATION OF TRAINING GOALS INTO TRAINING CONTENT

According to the principles laid down for this project (see Fig. 4), a description of training goals should state on the one hand what a course member should be able to accomplish after the training, and on the other what knowledge, skills and attitudes the students should acquire, bearing these desired future accomplishments in mind. The accomplishments are called 'terminal behaviour'. Knowledge, skills and attitudes are called 'readiness'. According to the approach chosen here, a description of training content should partly state the subject areas within which the students participating in the training should study and do exercises, partly give reference lists of different forms of documented study material and exercises within these subject areas. A possible technique for transforming training goals expressed in terms of behaviour into descriptions of readiness follows here. Those interviewed are given a situation described as a training goal and posed the question: 'What knowledge, skills and attitudes can be considered necessary for an employee to be able to deal adequately with the situation?' They are asked to go through all the situations that have been extracted as being of interest for the training in this way. This procedure can produce a list of descriptions of readiness linked to the situations directly related to the training.

In principle the technique for transforming the descriptions of readiness into descriptions of training content can be drawn up in the same way as the transformation mentioned above of situations of interest for the training into descriptions of readiness. In this case, those assessing the situations could start from a description of readiness and ask themselves: 'Which subject area(s) should the students study in order to acquire the desired degree of readiness?'

## FINAL COMMENTS

At present industry is entering a phase of semi-automatic work. During recent years the traditional division of labour has sometimes proved an obstacle to development in different ways. The work done at a particular place of work has traditionally been divided between different employees, who have not been allowed to carry out professionally work other than that which is considered to be part of a certain definite job. At the same time there is an increasing opinion demanding that monotonous jobs should be given more variety. Such a reform is desirable since making a job less monotonous gives a corresponding increase in satisfaction. One way of solving this dilemma would to be focus the analysis of needs on the places of work instead of analysing the training needs for the different jobs. This means that instead of carrying out job analyses one would do work place analyses wholly independent of the present division of tasks between different personnel. Once the tasks to be dealt with at the work places and the training needed for them have been established, jobs can be constructed on which claims other than the traditional ones can be made. An essential point in this context is naturally that, when the new tasks are constructed, the demands for high production efficiency and for a feeling of satisfaction with the job should both be met.

# La planification des programmes de formation à l'aide des méthodes d'analyse des emplois (résumé)

Kurt E. G. Gestrelius

Cette étude a pour base des travaux qui se poursuivent actuellement en Suède sur la formation des enseignants, mais qui sont applicables à la formation en général. L'auteur analyse des concepts comme ceux d'"analyse des emplois", d'"analyse du groupe optimal" et de "rentabilité de la formation", et présente un modèle des processus de planification de la formation professionnelle. Il souligne l'importance de la sélection et de l'évaluation des données nécessaires pour l'analyse des emplois, et expose le problème du passage des objectifs de la formation au contenu de la formation. Ses dernières observations portent sur la possibilité de modifier la division traditionnelle du travail, afin de répondre à une double exigence: assurer une productivité élevée et faire en sorte que les travailleurs soient satisfaits de leur emploi.

# Discussion

A speaker remarked that in planning programmes it was important to allow for differences in both ability and seniority. If one used a case-study approach one found that response to situations with which participants were confronted often varied according to their age and thus the courses had to be planned with these differences in mind.

The following participated in the discussion on 22 August 1972: Messrs Ter-Davtian, Bose, Tassios, Bureau, Ham, Krapfenbauer, Palmgren, O'Donnell, Barry, Niemi, Rink, Polke, Warner, Bertrand, Tymowski, Profos, Harboe and King.

# Modern information services in continuing education

Elin Törnudd, M.Sc. Chem. Eng.

## SYNOPSIS

Computer-based information services have been developed to cope with the threatening information explosion. Selective dissemination of information and retrospective searches by means of automatic data-processing methods are new services to be included in programmes of continuing education, both as a tool in engineering education and as a problem area not taught in pre-1970 curricula. Existing information systems are described and special emphasis is placed on data banks and interactive on-line systems. International networking is discussed, notably the Unesco programme UNISIST, which aims at the creation of a world-wide information system for science and technology.

Since the invention of movable type by Gutenberg, the application of computers to photo-typesetting is one of the most important innovations in printing technology. The information industry, which produces the abstract and index journals and other keys to primary information in science and technology has adopted the computer to be able to cope with the exponentially growing flow of information. This development has led to the availability of machine-readable files, which can be subscribed to instead of the printed version or as a supplement to the publication. These files stored on magnetic tapes can be used for two kinds of information services:

1. Selective dissemination of information, the continuing current-awareness service tailor-made for the individual user.
2. For retrospective searches of the literature on a given problem area.

The latter type of service is at present offered to a more limited extent, because it requires reorganization of the file to decrease the computer time needed.

## INFORMATION SERVICES AVAILABLE TO ENGINEERS

Information services should perhaps not be considered a direct means of continuing education of engineers, but rather a supporting function and a topic to be taught.

The main scientific and technological libraries and information centres are equipped to offer the following kinds of services to engineers and graduate students. The following examples have been selected from the 265 queries dealt with by the Helsinki University of Technology Library in May 1972 [1].

1. Quick reference questions put over the telephone or by telex and answered using manual methods.
   *Sample question:* What is the composition of the alloy Chromel Alumel?
2. Guidance of customers in the use of secondary information media is also carried out manually.
   *Sample question:* How should I go about finding a critical discussion of results published in 1966 in my dissertation on the theory of ladder networks?
3. Selection of material on a given topic which is carried out manually.
   *Sample question:* Send me copies of recent papers on environmental effect of gas pipelines.
4. Exhaustive retrospective literature searches can be carried out both manually and to an increasing degree through automatic data-processing methods.
   *Sample question:* What has been published since 1970 on nuclear reactor safety?
5. Selective dissemination of information—a current awareness service, usually computer based.
   *Sample question:* Supply me twice a month with a list of new references on the design and theory of optical systems.
6. Referral service carried out manually or using automatic data processing methods.
   *Sample question:* Which research workers in this country are presently engaged in research on exhaust-free motor cars?

## COMPUTER-BASED INFORMATION SYSTEMS

In accordance with the latest published inventory [2] prepared by the United Nations Economic Commission for Europe in co-operation with the Organization for Economic Co-operation and Development there were, in 1971, 100 different computer-based information systems, the majority of which cover different fields of engineering and allied sciences. Among these were: two systems in space sciences; eighteen in chemistry and chemical engineering; six concerning environmental science, building and transportation; seven in nuclear energy; six in physics; six covering a wide field of science and technology; and twelve on the specific technologies. Since 1971 the number of computer-based information systems has grown and increases continuously.

Some of the systems which we have been using for information services, are given below.

### *Multidisciplinary scientific and technical systems*

1. *Compendex.* Produced by Engineering Index, United States, is the computerized version of the *Engineering Index,* monthly, covering about 3,500 journals, conference proceedings and report series. The annual output is about 72,000 references from all fields of engineering.
2. *CICP—Current Index to Conference Papers in Engineering, Chemistry and Life Sciences.* Produced by the CCM Information Corporation, United States. The tapes include references to about 50,000 papers presented at conferences, accordingly the most recent information about research results and new developments.

3. *ISI–Science Citation Index Source Tape.* Produced by the Institute for Scientific Information, United States, covers from cover to cover about 2,500 journals and scientific serials from the fields of chemistry, biology, other sciences, medicine and technology. The number of references is 400,000 per year.
4. *Artikkel Indeks.* Produced by the Norwegian Center for Informatics in a Scandinavian co-operative effort, covers 800 journals from different fields of technology and industrial economics. The number of references is 22,000 a year.
5. *Bulletin Signaletique.* Produced by the Centre National de la Recherche Scientifique (CNRS), France, in several sections for different scientific disciplines.

*Chemistry and chemical technology*

6. *CAC–Chemical Abstracts Condensates.* Produced by the Chemical Abstracts Service, United States. The largest discipline-oriented service selecting from close to 15,000 journals and in addition from the book production and patents, about 250,000 references a year. Accordingly the coverage is the same as that of the journal *Chemical Abstracts.*
7. *FSTA–Food Science and Technology Abstracts.* Produced by the International Food Information Service, United Kingdom. The system covers books including bibliographies and journals including abstract journals from the field of food chemistry and food technology. The annual number of references is about 12,000.

*Physics and nuclear physics*

8. *Inspec–Information Service in Physics, Electro-Technology and Computers & Control.* Produced by the Institution of Electrical Engineers, United Kingdom, and the Institute of Electrical and Electronics Engineers, United States. *Inspec* contains about 120,000 references a year.
9. *NSA–Nuclear Science Abstracts.* Produced by the United States Atomic Energy Commission, has the same coverage as the corresponding abstract journal. The number of references is about 60,000 a year.
10. *INIS–International Nuclear Information System.* Produced by the International Atomic Energy Agency, Austria, through international division of input. *INIS* was started in May 1970 and grows annually with about 12,000 references to nuclear papers and reports.

*Metallurgy, mechanical engineering and transportation*

11. *Metadex–Metals Abstracts Index on Tape.* Produced by the American Society for Metals in co-operation with the Institute of Metals, United Kingdom. The system is the computerized version of the index of the abstract journal and contains about 25,000 references a year on the properties, processing, production engineering and use of metals and steel.
12. *MechEng–Mechanical Engineering KTH.* Produced by the Royal Institute of Technology Library, Sweden. The system covers about 200 technical journals with emphasis on production engineering. The annual number of references is 60,000.

13. *STAR–Scientific and Technical Aerospace Reports.* Produced by the National Aeronautic and Space Administration, United States. This system covers annually about 45,000 research reports.
14. *IAA–International Aerospace Abstracts.* Produced by the American Institute of Aeronautics and Astronautics, United States. The coverage is journal articles, patents and papers read at conferences, altogether 50,000 references a year.
15. *Artikkel Indeks for Skip.* Produced by the Norwegian Center for Informatics in a Scandinavian co-operative effort. The number of references on shipping and ship building in 1971 was 3,000.

*Forest products*

16. *Wood.* Produced by the Swedish Forest Products Research Laboratory and the Royal Institute of Technology Library, Sweden. The annual number of references is about 15,000.
17. *ABIPC.* Key-word supplement to the *Abstract Bulletin of the Institute of Paper Chemistry,* United States, is specially designed for retrospective searches and includes 10,000 references a year.

The above information systems are subscribed to and computer run by several scientific libraries and information centres in Scandinavia by the following centres (services number given in parentheses):

*Denmark:* Technical Library of Denmark (1, 6, 8, 13, 14).
*Finland:* Finnish Pulp and Paper Research Institute, Information Service (17); Helsinki University of Technology Library (10).
*Norway:* Norwegian Center for Informatics (4, 15).
*Sweden:* Biomedical Documentation Service of Karolinska Institutet (6 (and others)); Royal Institute of Technology Library (1-3, 6-9, 11-14, 16).

## USER APPROACH TO THE INFORMATION SYSTEMS

When an engineer wants to use the above listed systems for selective dissemination of information or for a retrospective search he is requested to formulate his field of interest in a few sentences and to supply some references to papers he recently has found to be of interest. He is also asked whether he wants a rather general search, which also carries along references to material of no interest, or a more strictly limited search, or a very specific search, which might lead to the result that even relevant references are missed.

On the basis of this information an interest profile is constructed by an information scientist in close co-operation with the customer (see Fig. 1).

A chemical engineer working in a plastics factory might provide the following description of his field of interest: 'I am interested in everything concerning polyvinyl chloride and especially in the destruction of chlorinated hydrocarbons in nature.'

For this question the following rather simple profile would be made up:

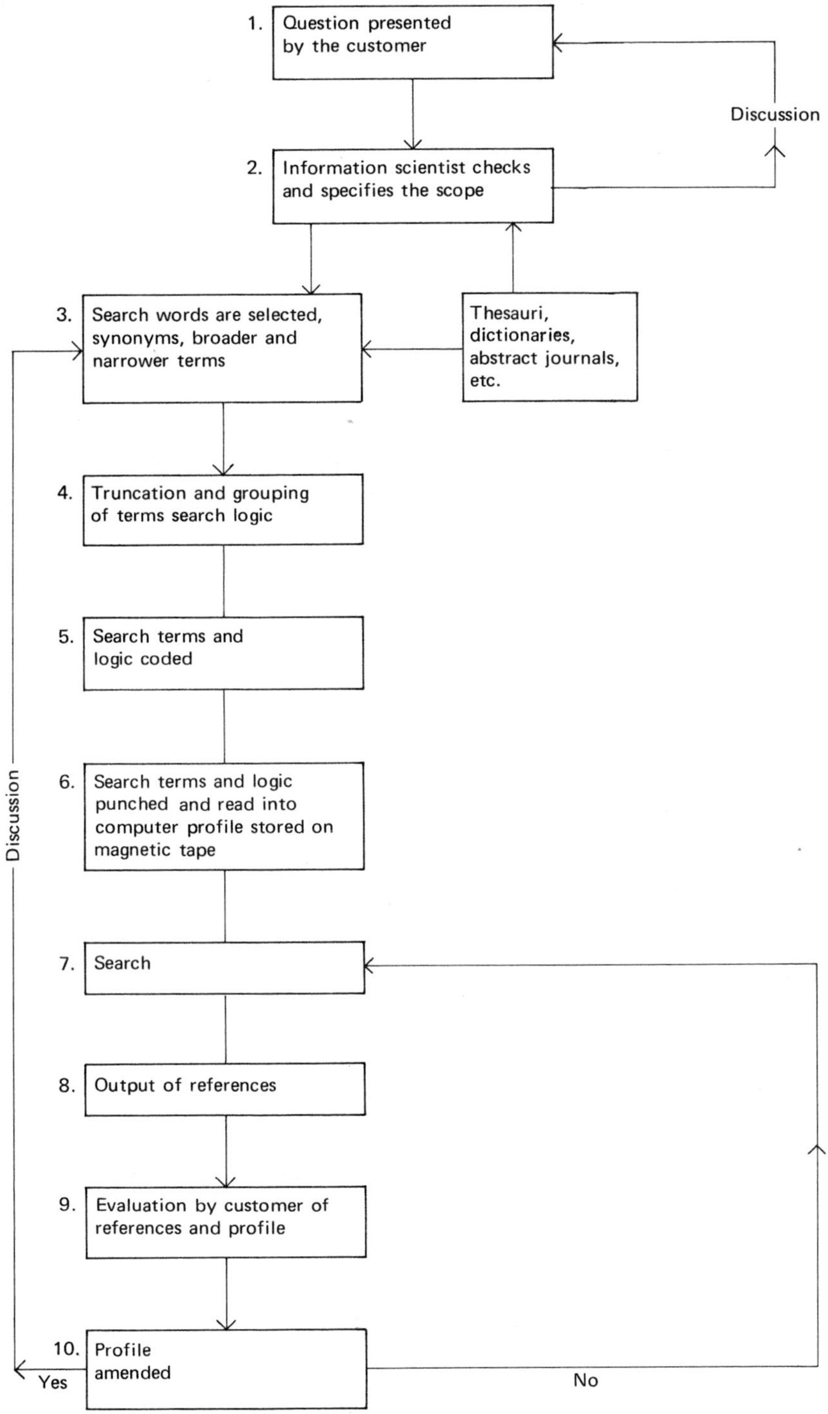


Fig. 1. *Construction and updating of profile for SDI.*

| Search term | | Search logic |
|---|---|---|
| A | PVC | |
| A | P V C | |
| A | VCM | |
| A | V C M | A |
| A | /polyvinyl chlorid/ | |
| A | /polyvinylchlorid/ | or |
| B | /chlor/ | |
| C | destruct/ | B and C |
| C | disposal | |
| C | waste/ | |

The slash at the beginning or at the end of a term means that the term has been truncated, and that words starting with a prefix or ending for example with a flexion will be taken into account. Accordingly the term /chlor/ above will pick up terms like dichloride and chloroform when matched against the biweekly data base on magnetic tape. This profile will accomplish a selection of references to all

GE)) TESTING OF PIGMENTS IN PLASTICS – TESTING OF PIGMENTS IN
NON-RIGID POLYVINYL-CHLORIDE (NON-RIGID PVC), PREPARATION OF TEST
SPECIMEN 1

KAUT GUM KU 24 359 71 N R N7 K5031
* PVC * VINYL CHLORID*

GE)) TESTING OF PIGMENTS IN PLASTICS – TESTING OF PIGMENTS IN
NON-RIGID POLYVINYL-CHLORIDE (NON-RIGID PVC) – PREPARATION OF BASIC
MIXTURE 2

KAUT GUM KU 24 358 71 N R N7 K5031
* PVC * VINYL CHLORID*

SYNTHESES OF ULTRAVIOLET-LIGHT ABSORBERS FOR POLYVINYLCHLORIDE (PVC)
. 1 .
GHATGE ND VERNEKAR SP 3
ANGEW MAKRO 20 165 71 004R NNOV L536
* PVC * VINYLCHLORID*

continued. . .

Fig. 2. *A sample of the print-out.*

papers on polyvinyl chloride and to those including the destruction or waste treatment of substances, the names of which contain the truncated stem /chlor/. (See also Fig. 2.)

After having received the list of tailor-made references in two copies the engineer marks those references of which he wants the original paper copied, and he receives the copies from the library within a couple of days.

I have gone into details in describing the SDI procedures, because the same method is used for retrospective searches disregarding the mode of use, on-line or off-line.

## COSTS OF COMPUTER-BASED SELECTIVE DISSEMINATION OF INFORMATION

Traditionally, library services have been offered free of charge to customers. Thorough retrospective literature searches and selective dissemination of information, however, go far beyond the classical library service and are usually at least to some extent paid for by the customer.

The usual price for an annual subscription of selective dissemination of information in Scandinavia is U.S.$75. This payment covers the computer costs for selective dissemination of information, provided that the number of customers per data base is large enough. Personnel costs and subscription of computer readable data bases are not covered by this price.

Since the use of computer-based services has not earlier been taught at the universities of technology, and since some of the services still are at an experimental stage, the policy has been to subsidize the service with government grants. Continuing education of engineers in order to give an insight into modern information services is accordingly supported by public authorities in several European countries, among them in all Scandinavian countries.

## INTERACTIVE ON-LINE NETWORKS FOR RETROSPECTIVE SEARCHES

The latest development is the creation of information-retrieval networks consisting of large information data banks accessible by telephone lines. These networks are especially suitable for retrospective searches answering questions like: 'What has been published on gas centrifuges since 1968?' Europe's first information retrieval network [3, 4] is that of the European Space Research Organization (ESRO).

This network consists of a central information data bank in the ESRO computing centre in Darmstadt and remote consoles in libraries and information centres in Paris, Rome, London and in two Scandinavian cities. The Technical Library of Denmark and the Royal Institute of Technology Library in Stockholm installed their telephone line to Darmstadt on 1 June 1972. The ESRO information data bank orginally contained the NASA files from the United States which are updated and supplemented with the information from several available computer-based systems, such as the above-mentioned *Compendex* and *Nuclear Science Abstracts.*

The engineer performing the search uses a remote console consisting of a television screen and a keyboard. He may enter his inquiry in plain language through the keyboard on to the screen and transmit his message by telephone line to the computer in Darmstadt. Attached to the screen and keyboard is a teleprinter which records the dialogue carried out. In the course of the dialogue with the computer the engineer can modify his query in the light of information supplied to him on the screen by the computer. After having made up his mind in regard to the best possible search profile constructed similarly to that described above in connexion with selective dissemination of information, he may ask the computer to give the full reference to some of the documents satisfying his search profile for review. Finally he may command all references to be printed on-line by the tele-typesetter at his side or he may ask for off-line printing by the line printer in Darmstadt. The latter procedure causes a slight delay, since the output is mailed. It is, however, more economical, if the number of references is large.

## FUTURE OF INTERNATIONAL NETWORKING AND THE UNISIST PROGRAMME

Unesco and the International Council of Scientific Unions (ICSU) have carried out a study on the feasibility of a world science information system (UNISIST). An intergovernmental conference was organized by Unesco in October 1971, the recommendations of which will be decided on by the Unesco General Conference in November 1972 [5].

The proposal is to launch a long-term international programme on transfer of scientific and technical information with the following objectives:

To advance and co-ordinate the world trends towards information-sharing and co-operative agreements among governments, international organizations and operating information services.

To provide guidance and catalytic action for the necessary development in the field of scientific and technical information.

To facilitate the access to published information by scientists, engineers and technologists.

To help the developing countries meet their needs for scientific and technical information.

To take the necessary actions for the establishment of a flexible network of information systems and services based on voluntary co-operation.

In my opinion one of the most important targets of the UNISIST programme will be to co-ordinate the existing and planned information systems for different disciplines and missions. A second important target is to make it possible for engineers in both industrialized and developing countries to gain access to the systems and to get an insight into their use.

It has been claimed that computer-based information systems are too expensive to be made available to engineers in developing countries and small industrialized countries. In my opinion this argument is incorrect. Computer-based information services are expensive, that is true, but their unit cost decreases with the number of customers. Accordingly a system set up for a certain country or region can be used by others at marginal costs.

International co-operation in modern information services is accordingly a necessity, if we want to offer continuing education in the form of a current awareness service or in the form of retrospective searches on findings made till the present time. The UNISIST programme will, without doubt, play an important role in the continuing education of engineers all over the world.

*References*

1. 'Educating the Library User', *Proceedings of Fourth Triennial Meeting of the International Association of Technical University Librairies,* Loughborough, 1970.
2. *Inventory of Major Information Systems and Services in Science and Technology,* Paris, Organization for Economic Co-operation and Development, 1971.
3. Björn V. Tell and Stig Fröderström, 'Information System for Design Engineering', *Management Informatics,* Vol. 1, No. 1, 1972, p. 10-18.
4. N. E. C. Isotta, 'Europe's First Information Retrieval Network', *ESRO/ELDO Bulletin,* No. 9, 1970.
5. Unesco, *UNISIST. Intergovernmental Conference for the Establishment of a World Science Information System. Final report,* Unesco, Paris, 1971.

# Services modernes d'information en matière d'éducation permanente (résumé)

Elin Törnudd, M. Sc., ingénieur chimiste

Des services d'information sur ordinateur ont été réalisés pour faire face à la croissance exponentielle de l'information. La diffusion sélective de l'information et les recherches rétrospectives par des méthodes de traitement automatique des données figurent parmi les nouveaux services à inclure dans les programmes d'éducation permanente, car il s'agit là d'un moyen de formation des ingénieurs et d'une question ne figurant dans aucun programme d'enseignement avant 1970. L'auteur décrit les systèmes d'information actuels, en insistant particulièrement sur les banques de données et les systèmes en ligne réagissant l'un sur l'autre. Il examine la question de l'établissement de réseaux internationaux et notamment le programme UNISIST de l'Unesco, qui vise à créer un système mondial d'information scientifique et technologique.

# Application of technology in continuing education

Monroe W. Kriegel

## GENERAL COMMENTS

Many interesting and exciting technological developments have been made in education in the United States, both at the secondary-school level and in colleges and universities. Among these have been the use of television, both live and taped; the application of the Electrowriter, or 'blackboard by wire'; extensive use of recorded lectures or audio systems, both through the use of tapes and live telephone lectures; the development of programmed instruction texts; the use of multiplex cable for transmission of television signals; and the use of satellites for transmitting frequency modulation (FM) and television (TV) signals. These various approaches will be examined with regard to the procedures followed, the effectiveness of each as a method of presenting new material and the cost factors involved.

## USE OF LIVE TELEVISION

There are at present four major university instructional television (ITV) networks. These are the GENESYS system in Florida, the TAGER operation in north Texas, the Stanford system in California and the Higher Education Televised Instruction Network System in Oklahoma. In addition, smaller scale versions of similar systems are in use in Minnesota, Wisconsin, Ohio, Rhode Island, Michigan, and Nebraska.

There are a multitude of TV transmission means of linking institutions with off-campus students. These fall into two general categories: area coverage systems and point-to-point systems. The area coverage systems include Very High Frequency-Ultra High Frequency (VHF-UHF) (see Fig. 1), broadcast TV, community antenna TV (commonly called CATV), instructional television fixed service (ITFS), and satellites. Point-to-point systems include use of common carriers such as telephone line, private microwave, ITFS and cable.

Essentially all educational television is in black and white rather than colour. This is due to the high cost of colour cameras and the production skill necessary to obtain high-quality film. Also tests have shown that colour is not necessary to enhance the learning process. In other words, 'better a good quality black and white than a poor quality high-cost colour'.

*Area coverage instructional television systems*

*VHF-UHF*

The earliest area coverage system for educational purposes used standard VHF-UHF TV transmission (see Fig. 1). Broadcasts were played over both commercial television stations and special educational channels. This method was quite popular in the late 1950s and early 1960s. Educational programmes are still being broadcast over the conventional channels in the early morning hours between 6 and 7 a.m. and over educational television channels in the early evening. However, most of the educational offerings are now in the areas of the humanities. Essentially nothing is available in the fields of science and engineering. This mode of transmission reaches thousands of students with one-way TV. Talk-back is impractical because of the large number of students in any one class. The VHF-UHF systems are not widely used for university instructional TV systems. Capital and operating costs are relatively high. Capital costs range between $250,000 to over $1 million per channel. Annual operating costs per channel range from about $120,000 for a minimum operation to over $400,000 for a professional operation.

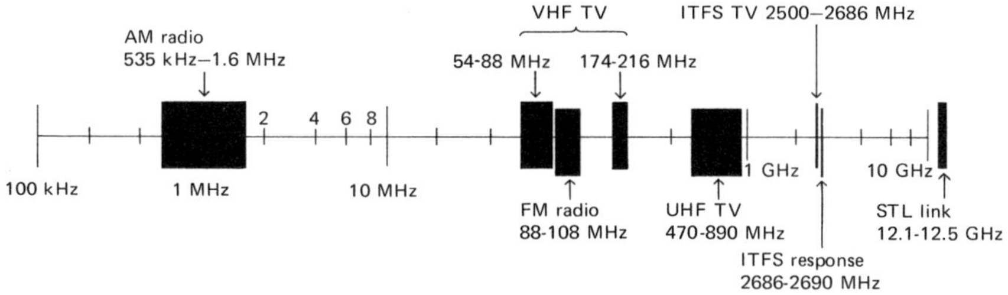


Fig. 1. *Spectral distribution of broadcast services* [*3*].

*ITFS*

The most common type of instructional television is the fixed service, known as ITFS. In 1963, the Federal Communication Commission allocated thirty-one ITFS channels [1] in the 2500 to 2686 MHz band (see Fig. 1). By July 1969, there were applications for 174 channels, and thirty-eight channels actually in use. As of January 1971, there were sixty-five ITFS systems and 157 channels operational. However only a handful of these were allocated to institutions of higher education. These channels use standard TV format and are typically allocated in groups of four. Maximum transmitter power output is restricted to 10 watts per channel. However, the area of coverage can be considerable. Tables 1 and 2 below [1] present annual operating budget and capital cost for a typical university ITV system using ITFS TV transmission. Typical instructional ITV university systems show operating costs to be between $10 and $30 per channel per hour.

It may be of interest to compare briefly the four major operating TV systems in the United States.

*Table 1.* University ITV system annual operating budget [1]

| | Cost in dollars per number of channels assuming 8 hours/day/channel | | | | Notes |
|---|---|---|---|---|---|
| | 1 channel | 2 channels | 3 channels | 4 channels | |
| *Fixed* | | | | | |
| Director | 9,000 | 9,000 | 18,000 | 18,000 | Assume half time for 1 and 2 channels |
| Secretary | 0 | 0 | 3,000 | 6,000 | Assume school supplies part-time for 1 and 2 channels, half time for 3 channels |
| Chief engineer | 16,000 | 16,000 | 16,000 | 16,000 | |
| Engineer | 0 | 5,000 | 10,000 | 10,000 | Half time for 2 channels, full for 3rd and 4th channels |
| Engineer (2nd) | 0 | 0 | 0 | 5,000 | Half time from school |
| Driver | 2,000 | 2,000 | 4,000 | 4,000 | Half time for 1 and 2 channels ; full time for 3 and 4 channels |
| *Direct labour* | 27,000 | 32,000 | 51,000 | 59,000 | |
| Overhead at 20 % | 5,400 | 6,400 | 10,200 | 11,800 | |
| Car | 1,000 | 1 000 | 2,000 | 2,000 | |
| Tower rental | 5,000 | 5,000 | 5,000 | 5,000 | |
| *Fixed costs* | 38,400 | 44,400 | 68,200 | 77,800 | |
| *Variable* | | | | | |
| Operator costs at $2.50 per hour | 3,500 | | | | 1,400 hours, 1 channel[1] |
| | | 7,000 | | | 2,800 hours, 2 channels[1] |
| | | | 10,500 | | 4,200 hours, 3 channels[1] |
| | | | | 14,000 | 5,600 hours, 4 channels[1] |
| *Total operating cost* | 41,900 | 51,400 | 78,700 | 91,800 | |
| Cost/hour exact | $ 29.90 | $ 18.30 | $ 18.75 | $ 16.40 | |

1. Assume 8 hours/day/channel, 5 days/week, 35 weeks/year.

*Source:* GENESYS Systems Inc. 1054 East Meadow Circle, Palo Alto, California 94303 (United States).

*GENESYS system in Florida.* The letters stand for Graduate Engineering Education System [2]. The system was established in 1963 by an act of the Florida legislature. Figure 2 shows the geographic scope of GENESYS. Operation began in April 1965 with studio classrooms at the main campus in Gainesville and at 20-acre satellite campuses at Daytona Beach, Orlando, and Port Canaveral,

*Table 2.* University ITV system capital budget (in United States dollars) [1]

| | Number of ITFS channels | | | |
|---|---|---|---|---|
| | 1 | 2 | 3 | 4 |
| Consulting and legal fees | 20,000 | 20,000 | 20,000 | 20,000 |
| Programme management, design engineering and drawings | 35,000 | 35,000 | 35,000 | 35,000 |
| Installation and test | 44,000 | 57,000 | 69,000 | 80,000 |
| Studio classrooms equipment | 23,000 | 46,000 | 69,000 | 92,000 |
| Studio control | 19,000 | 37,000 | 56,000 | 74,000 |
| Master control | 8,000 | 27,000 | 46,000 | 49,000 |
| RF transmission equipment emergency power | 75,000 | 88,000 | 102,000 | 116,000 |
| Talk-back receiving equipment | 26,000 | 27,000 | 29,000 | 30,000 |
| Spare parts | 6,000 | 12,000 | 18,000 | 24,000 |
| Test equipment | 15,000 | 15,000 | 15,000 | 15,000 |
| Totals | 271,000 | 364,000 | 459,000 | 535,000 |

*Source:* GENESYS Systems Inc., 1054 East Meadow Circle, Palo Alto, California 94303 (United States).

just south of Cape Kennedy. Full-time professors reside permanently at the satellite campuses and courses originate from all four campuses. The network regularly carries from forty to fifty graduate courses, mainly in electrical, systems and aerospace engineering, with fewer offerings available in mechanical, civil, chemical engineering and mechanics. Graduate courses for minor credit are also given in mathematics, statistics and physics. Credit courses are scheduled from 6.30 a.m. to 9.55 p.m., Monday to Thursday with Friday and Saturday reserved for non-credit short courses, seminars and oral examinations.

Students may participate from any of the four studio sites as well as in three auxiliary classrooms. The students are provided microphones for communication with the professor and with other students. This important feedback audio system, known as 'talk-back TV' permits all students in the GENESYS system to hear questions from all campuses. Students at one location may argue not only with the professor but with students at other locations.

Early in 1966 a programme of non-credit short courses was initiated for transmission on Friday and Saturday when graduate courses were not scheduled. Thus the entire technical community, including management, was able to benefit from talent available over the TV system in the professional improvement programme. Some short courses have enrolled as many as sixty participants resulting, however, in reduced talk-back possibilities.

The GENESYS system is still in operation, but due to the reduced aerospace programme in the Florida area, the number of participants has greatly decreased. Recently the system was taken over by the state and plans are to expand the offerings to the area of teacher education and the humanities.

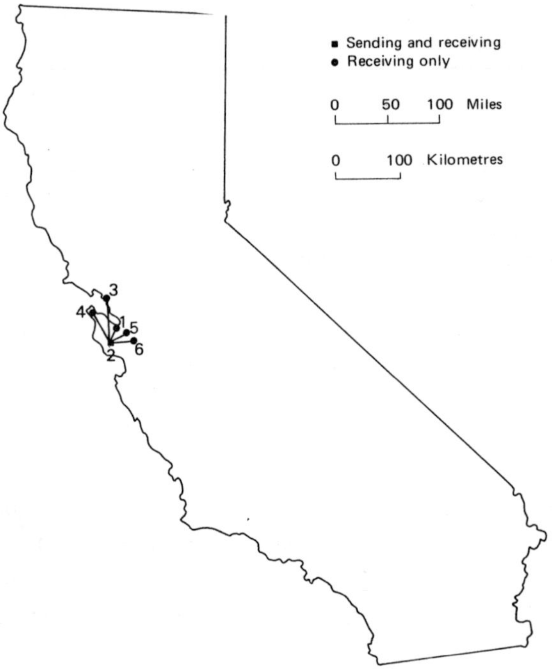


Fig. 2. *GENESYS system (Florida). 1. Gainesville, University of Florida; 2. Daytona Beach; 3. Orlando; 4. Cape Kennedy; 5. Patrick Air Force Base; 6. Kennedy Space Center; 7. Naval Training Device Center; 8. West Palm Beach; 9. Boca Raton; 10. Eglin Air Force Base; 11. University of West Florida; 12. University of South Florida.*

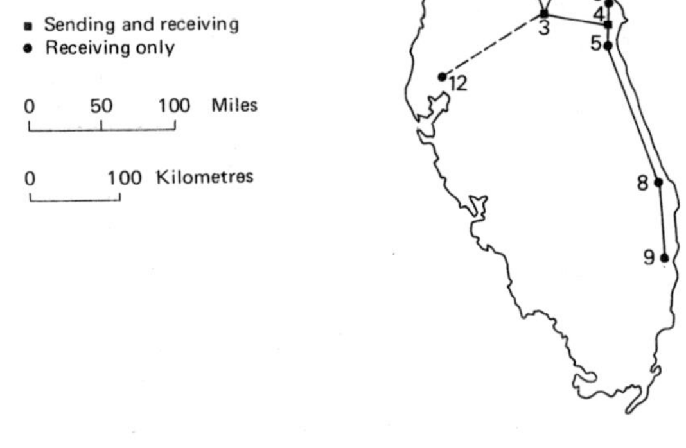


Fig. 3. *Stanford system (California). 1. Stanford University; 2. Black Mountain Transmitter; 3. University of California-Berkeley; 4. San Francisco; 5. University of Santa Clara; 6. San Jose State University.*

*TAGER system in north Texas.* The second major educational television network in the United States is known as TAGER, which stands for The Association for Graduate Education and Research. It began in August 1965 when seven north Texas private educational institutions announced the pooling of their physical and academic resources in order to enhance graduate education in the Dallas-Fort Worth region of north Texas. Similar in nature to the GENESYS system, TAGER is talk-back TV linking together classrooms and studios in Austin College, Bishop College, Dallas Baptist College, Southwestern Medical School, Texas Christian University, Texas Wesleyan College, the University of Texas at Dallas and the University of Dallas, with industrial classrooms in thirteen company plants. The system began operation in January 1967. Classes are taught continuously from 8.00 a.m. to 8.00 p.m., Monday to Friday, with broadcast studios on the campuses of the universities and over fifty classrooms in industrial sites in the area. As of spring 1972, TAGER has offered 700 courses to approximately 15,700 students on nine campuses and in eleven industrial plants. Enrolment for the current semester is around 2,000.

Industrial students taking a course for credit by a TAGER television system, either in a company classroom or in a school classroom which is remote from the original classroom, pay $60 per semester credit hour course charge, $40 per semester credit hour television surcharge, or a total of $300 total course charge for a three-semester hour course. The fee for auditing a course is $150.

*The Stanford Instructional Television Network.* The Stanford ITV network [3] operating in the general area of San Francisco, connects Stanford University with the University of California at Berkeley and reaches somes sixteen charter member companies in the San Francisco Bay area (Fig. 3). Co-operative arrangements with San Jose State University and the University of Santa Clara are also under consideration. This is another talk-back TV system utilizing live television broadcasts. Two channels of the television system were initiated in April 1969, with fifteen graduate engineering courses and two seminars being broadcast on a regular basis. In October 1969, the remaining two channels were implemented and a total of thirty-one Stanford courses representing all ten departments in the School of Engineering were televised. During the winter quarter of 1972, a total of 378 courses were offered with a total attendance of 2,849 plus 532 auditors. These have covered all fields of engineering, with the largest number in electrical engineering. At present, twenty-seven organizations are members of the network. Of these, seventeen are equipped to receive televised instruction in sixty-three separate classrooms.

The total cost of the Stanford portion of the four-channel network was about $625,000. This included all broadcast and transmission facilities. This was borne largely by the companies, with each participant's share a function of its size, and could be paid either in a lump sum or pro-rated over five to ten years. The participants were responsible for installing their own receiving facilities subject to technical specifications in guidelines provided by Stanford.

Network member organizations are permitted to make 'off-the-air' video tapes of Stanford lectures at the request of their students who wish to make up missed classes or use them for course review. Such tapes cannot be employed in any other way and must be erased within two weeks of the end of each quarter. Tapes are also occasionally made at Stanford for the same purposes or at the instructor's request for study in self-appraisal.

One option which may be of interest was the creation of a 'non-registered option' which permits industry graduate students to take televised courses in remote classrooms without being matriculated at the university. They are tested and graded with the same standards as regular students but their performance does not affect university standards. Included in this category are students who already have degrees, students who do not want to complete all the work for a degree, and students who want a degree but are not at present academically qualified. For this last category, Stanford will allow such students to become matriculated and to 'transfer' completed units by petition, if as non-registered students they perform well enough in competition with regularly enrolled students. Auditors from participating organizations are permitted at remote classrooms for televised courses at reduced fees. They receive no grading or testing, nor are records kept for them. Also selected seminars in research fields of interest to the network members are televised and are available to all personnel from these organizations at no cost.

A separate non-profit corporation, the Association for Continuing Education (ACE) was established to provide programming of non-credit courses of particular interest to industry. ACE has a full-time general manager who works with a curriculum committee from the companies in developing courses and hiring instructors. The courses include testing and grading and a certificate issued to each student and to his employer upon completion of a course. ACE operates on a break-even basis, recovering its cost from nominal tuition charges. The courses are also televised live to remote students in the San Francisco Bay area who are given the same talk-back capability that is provided for Stanford students.

*Oklahoma Higher Education Televised Instruction System.* The most recent of the university televised instruction network systems is that established in the State of Oklahoma. It differs from the others in that it links together the major universities of this state and the state medical school with major industrial organizations and some junior colleges. It is an example of a private point-to-point microwave system servicing several ITFS area systems, as shown in Fig. 4. A microwave relay system transmits televised courses from four campuses to eight cities. The central control tower is located at the University of Oklahoma Medical Center in Oklahoma City. Courses are broadcast from the Medical Center, University of Oklahoma campus at Norman, Oklahoma State University at Stillwater and the University of Tulsa at Tulsa. Closed circuit two-channel ITV courses are received in all sending stations plus Muskogee, Bartlesville, Ponca City, Enid, Duncan and Ardmore.

The Oklahoma system was authorized by the 1970 Oklahoma legislature. The nineteen hop microwave contract was awarded for $888,835. The total approximate cost including the ITFS transmission equipment and the studio classrooms is $1.7 million. The amount of state funds authorized was $1 million with the remainder to be collected from industrial organizations on the basis of $1,000 per 'seat'. Each seat entitles the company to enrol one student each semester. A contribution of $25,000 permits unlimited enrolment. In addition each company is responsible for purchasing its own receiving equipment and equipping its classroom. Tuition was set at $75 per credit hour with the institution offering the course to receive its normal tuition and the remainder to go to the Oklahoma Regents for Higher Education, which operates the system. For example, at Oklahoma State University (OSU), the normal campus tuition for in-state students is $14 per credit hour. Thus each industrial student enrolled in a

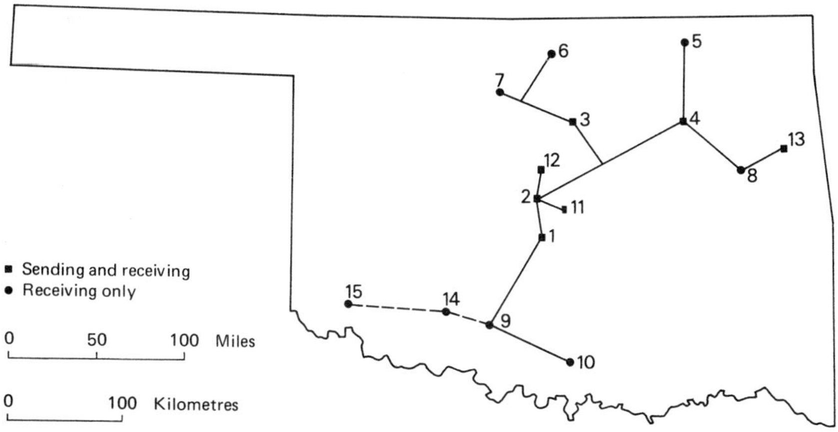


Fig. 4. *State of Oklahoma system.* Original: *1. University of Oklahoma-Norman; 2. Oklahoma Medical School-Oklahoma City; 3. Oklahoma State University-Stillwater; 4. University of Tulsa-Tulsa; 5. Bartlesville; 6. Ponca City; 7. Enid; 8. Muskogee; 9. Duncan; 10. Ardmore. To be added (autumn, 1973); 11. Oscar Rose Jr. College-Midwest City; 12. Central State University-Edmond; 13. Northeastern University-Tahlequah; 14. Lawton; 15. Altus Air Force Base.*

course originating at OSU, pays the equivalent of $14 to OSU and $61 to the State Regents for each credit hour. Students at other universities which are part of the system pay only their normal campus tuition.

The system was completed and used on a trial basis in the mid-spring of 1971, with the first full programme in the autumn. In spring 1972, the TV programme offered seventy-two courses, including thirty-three in engineering, thirteen in chemistry, and nine in mathematics and computer science.

In order to achieve wider coverage, additional receiving outlets will be established in autumn 1973 in five additional places, of which three will be equipped to transmit (Fig. 4). It is anticipated that within the next few years the majority of the junior colleges in the state together with military bases, governmental units and additional industries will be linked to this system.

The operation of the Oklahoma system is essentially the same as that in the other states. Currently there is a single classroom studio at each originating point. In the back of the room, housed behind a glass panel, is the operator for two cameras. One camera focuses on the speaker who sits at a table in front of the room and the other is directly overhead and focuses on the writing pad. The operator can project from either camera or from both simultaneously, thus giving a split image on the screen.

During the spring of 1972, two non-credit courses for adults in distant cities were tried on an experimental basis. The first of these was a course in 'Understanding the National Electrical Code' offered on four successive Tuesday evenings, each session lasting three hours. There were 147 attendees in ten locations scattered over the state. Most of these were electrical contractors or technicians ranging in background from high-school training to one or two years of college. The course was very well received and the participants found that they could learn and could actually participate in the discussion in this type of teaching. This was our first experience in the use of television as a means of teaching a practical course to a large number of technicians. The results were most gratifying.

The second non-credit offering was a course in review of accounting for candidates for the certified public accounting examinations, taught to thirteen locations over the state. The review was presented in two parts: auditing and accounting practice. The auditing review sessions were telecast from 1 p.m. to 5 p.m. on consecutive Saturdays from 12 February to 4 March to 115 students, and the accounting practice review sessions from 1 p.m. to 5 p.m. on consecutive Saturdays from 11 March to 29 April to 109 students.

*On-campus closed-circuit television*

A large number of universities are using an on-campus closed-circuit television network system in order to have a single instructor teach many sections of a class simultaneously. This system of teaching has a number of advantages. Through the use of multiple receiving units, students are able to see the face of the lecturer much more easily than they could in a large classroom in which there might be as many as 1,000 students. Of more particular advantage, however, is the use of 'close-ups' to bring the details of demonstration experiments simultaneously to large groups. This is particularly helpful in those classes in which techniques are taught. A chemistry professor can demonstrate correct laboratory technique in a much better way, because all the students can see easily. In the teaching of drafting, the proper use of the drawing pen and auxiliary equipment can be demonstrated easily. The students get more of a 'feel' for the professor they see on TV than they do of the faint image they can barely see in the front of a large lecture hall and whose voice comes to them only through strategically placed speakers.

## USE OF TELEVISION TAPE FOR INSTRUCTIONAL PURPOSES

For a number of years there has been a growing use of television tapes for instructional purposes in the secondary school system. These are used largely to supplement the normal teacher-lecture type of instruction. Customarily, the tapes are used to illustrate a particular segment of the subject being taught. Some schools prepare their own tapes and supplement these with purchased tapes which are kept in a central library. The use of TV in secondary schools is now so widespread that there will be no attempt to describe it fully in this article.

Three universities are now offering an extensive programme off campus through the use of videotapes made under actual classroom conditions and then sent out by courier or by the United States Postal Service to off-campus classrooms. These are Iowa State University, Colorado State University, and the University of Colorado.

*Iowa State University Videotape Program*

The Iowa State University started with a pilot course in industrial engineering on 'statistics for engineers' in the spring quarter of 1968. Results were encouraging and the programme was begun in earnest in the spring quarter of 1969. The 1969/70 academic year was the first full year of operation, in which some 300 different men participated in at least one course among the thirty-two credit offerings, and one refresher course for working engineers wishing to take the 'engineer-in-training qualifying examination'. The capability of making duplicate tapes allowed the university to teach forty-one different sections in seven cities over the academic year. Some twenty-six different instructors from eight separate academic departments were involved in the effort. Employees from twenty companies made up the student body.

Three classrooms within the college of engineering are equipped for videotape classes. Each has one camera, two recorders and related equipment. The videotape programme has priority in scheduling classes in these rooms. Each classroom has at least two cord microphones in addition to one for the instructor to facilitate recording questions and answers in class discussions. A student cameraman is expected to produce a good tape or tapes by noon of the day following the class meeting; this allows time to remake a tape if necessary.

The 'engineer-in-training refresher non-credit course' is handled somewhat differently since it is not connected with any on-campus course. Iowa State has, and permanently maintains, several sets of tapes for this course. These would normally be shown in the autumn or winter prior to EIT examinations in December and April. Although the course was developed with engineering graduates in mind, some non-engineering graduates have taken the course and later passed the EIT examination. A flat fee is charged for a one-time use of these tapes regardless of the number of people who have taken the course.

Each location off campus appoints a site co-ordinator. This person directs the videotape activities at the site such as ascertaining which courses their people might want, registering students for the courses, arranging for classrooms and equipment and receiving, showing and returning tapes. The playback equipment, which can be leased or purchased, is provided by the industrial organization. Tapes are usually received at the site about two days after being recorded and are shown approximately a week later. Normally, the off-campus classes meet regularly and with the same frequency as the on-campus classes. For example, if the on-campus class meets for fifty minutes three times a week, the normal procedure is for the off-campus class to do the same. Course instructors are encouraged to visit each off-campus class at least twice during the course. In addition, arrangements are made for the instructor to be in his office at certain times to receive calls.

In general, the off-campus students appear to learn as much from the tapes as those on campus participating in the live presentation. Results of examinations are essentially the same for students having roughly the same background. The professors get no additional pay for participating in the television programme although usually they feel the obligation to do a better job of preparation and presentation. Normally, tapes are erased when they are received back at the campus, so there is no question of royalty payment to the instructor.

*Colorado State University Videotape Program*

For the past five years, Colorado State University has provided complete graduate study programmes for engineers employed in Colorado's industrial and government facilities. Under the Colorado State University Resources in Graduate Education (SURGE) programme, regular campus classes add part-time students who participate by seeing the classes on videotape on a two-day delayed schedule at their place of employment. The faculty maintain consultation hours by telephone and visit the industrial sections twice a quarter. The videotaping technique has been extended to provide educational programmes for the students and teachers of community colleges and high schools, and also inmates of the state penitentiary. This programme is known as CO-TIE which is an acronym for Cooperation via Television Introduction in Education. An average of twenty-six courses per quarter were offered in 1971/72 to forty-seven off-campus locations and ninety-five student groups. Over 400 videotapes were delivered by Colorado State University each week of the academic year.

The videotape programme at Colorado State University is by far the largest in the United States, and is considered highly successful. Further information is available from Dr Lionel V. Baldwin, Dean of the College of Engineering, Colorado State University, Fort Collins, Colorado 80521.

*University of Colorado Videotape Program*

The University of Colorado has conducted a programme of videotaped engineering courses during the past four years. In most of these courses, actual classes on campus are videotaped; the tapes are then sent to various remote locations for viewing by students. This permits full-time employees of a variety of organizations to take university courses for academic credit without having to go to the campus.

Classes are videotaped in a specially equipped classroom with space for ninety-six students. A remotely controlled camera at the back of the room is used to tape blackboard discussions and an overhead camera is used to tape written material on the desk pad. The lecturer has desk controls to switch between cameras and to aim, focus and zoom. Duplicate controls are available to the recording technician so that either the lecturer or the technician can control the recorded picture. This procedure is essentially the same as that previously discussed for live television except that in the latter case the operation of the camera is handled entirely by the technician.

Generally it has been found that students learn as well from a videotape as they do from a live lecturer. The ease of learning is largely a function of the quality of presentation, the clarity of the picture and the distance from the student to the screem. For best results, no more than twenty students should view a single standard 24-inch (61 cm) monitor at one time. Ten students per monitor is better and the ideal is one or two students, with a single monitor and with the capability of stopping, starting and replaying the tape in accordance with their speed of learning.

One disadvantage noted is that it is more difficult to take notes from a small monitor than from a blackboard. In some courses this has been overcome by giving each student a prepared set of notes containing all the written material which he may need and which he would normally copy down during the class. Class questions are handled by having the instructor repeat the question into a recording microphone so that the videotape class will hear it. In some cases, microphones are actually suspended overhead in various places throughout the class in order that questions may be picked up and recorded.

It is necessary to have some contact between the professor and the students. This is usually handled by having the professor visit the class about once every five weeks. In addition to this, arrangements are made for telephone contact between the students and the instructor at regular intervals.

Student reaction has generally been favourable to the videotape programme. A serious disadvantage is the lack of easy access of the instructor for answering questions at the exact time an answer is needed for properly understanding the presented material. A major advantage is that industrial students who are out of town during a regularly scheduled playing of the videotapes may view them at their convenience at a later time. Also those students who wish may watch all or parts of lectures a second or third time if desired.

The faculty objections are the usual ones: concern with the fact that off-campus students are not exposed to the intellectual environment of the campus and with academic standards. In answer to the latter, a careful analysis of grade records of students taking the same course live on campus and comparing them with those of students taking it by videotape showed no appreciable difference in learning. In fact, grades in the off-campus courses were slightly higher, but the difference was considered to be statistically insignificant.

The cost of videotaping equipment and facilities can vary greatly, depending on the quality of equipment, number of classes, classroom furniture, etc. The minimum Sony half-inch tape equipment, including camera, tripod, monitor, carrying case, and other auxiliary equipment is about $2,000. The University of Colorado has estimated operational costs to be $1,630 per three-credit hour course. Colorado State University has made a very complete cost analysis based on its five years of operation and a level of 80 courses per year, and arrived at $32.25 per recording hour and $4.75 + $1.30 x student enrolment per delivered tape. This includes amortization of all equipment.

*Use of videotape by professional engineering societies*

A number of the professional engineering societies have produced both video- and audio-tapes for use in continuing engineering education of their members. Space does not permit discussion of all these, so by way of illustration a major programme undertaken by the Society of Petroleum Engineers, a branch of the American Institute of Mining Engineers, will be presented. The SPE videotape series began in 1967 when Dr Ben Caudle, head of the School of Petroleum Engineering at the University of Texas, was commissioned to develop a course on fundamentals of reservoir engineering. This 40-hour unit was prepared in a studio set up in the basement of his home, and was based on a similar non-credit course he had taught to local sections of the society. A follow-up course was developed by Dr Caudle in 1968. Response to these two courses was great enough to convince the society and full-scale production of videotape courses was started. A studio was constructed at SPE headquarters, complete with half-inch recorders, TV cameras, special effects generator, lights, etc. In June 1970, Dr Folkert Brons was brought to Dallas to tape a 13-hour course on statistics for petroleum engineers. At the same time the SPE Continuing Education Committee began developing a plan for ten short courses, each 10-15 hours in length, under the broad heading of petroleum operations. Each is to be taught in segments by individuals within the industry who are considered experts in particular areas of production operations. The first course – production operations course I: well completions – was divided into four segments; invitations were issued to four instructors who were then brought to Dallas for orientation. The men then returned home and spent several months preparing their presentations, as well as typed copy for the course notebook. Segments of the course were taped as each man completed his material, and the production work was finished in late 1971.

The course was announced to local sections in December 1971, and by February 1972 twelve local sections had been scheduled to take the courses during the spring months.

Production work and taping have begun on the second course in the series 'well diagnosis'; completion and scheduling by local sections is planned for January 1973. Future plans call for one additional course each year.

Over-all, since the programme began in September 1967, thirty-one of the society's sixty-two sections have participated in the programme with about 10 per cent of the 15,000 members taking at least one videotape course. The SPE charges its local sections $25 per student per course, and recommends a class size of from ten to twenty-five students. It has six play-back units which are loaned to local sections.

## VIDEO CASSETTES

The most convenient use of videotape is the form of video cassettes. Programmes are easy to record and play back, and the transportation problem is greatly simplified. However, the simultaneous development of videocassettes of various types by different companies caused problems which have only recently begun to clarify.

By way of background, the first recording period is known as the Ampex era, which used Helical scan machines from about 1962 to 1966. These initially cost about $10,000 each, but the Helical scan tape size dropped to a half-inch in 1965 and the price fell into the $1,000 to $3,000 range. Helical colour came on the market about two years later. However, in March 1966, the Columbia Broadcasting System announced the electronic video recording commonly called EVR, which used film. In January, 1970, the first EVR colour was developed. Problems developed when the Federal Communications Commission found that a Motorola EVR player gave off signals which might interfere with VHF or ordinary TV and UHF spectrums. Development is still continuing.

Another device called 'cartridge television' was announced in July 1971. The features of this system are two-track audio and two-hour maximum play time on its video cassette which is achieved through multiple scans. Cartridge television is aimed primarily at the consumer market and the system has been busy assembling a library of some 1,000 titles, some of which are training course material. The company will also market blank cartridges in 14-, 30-, 60- and 114-minute lengths, with the price of a blank hour-long unit estimated at about $16-$20. The cost of the single-chassis player and 19-inch (48 centimetre) colour TV set is projected at $900, and the free-standing player only unit at $500.

The first video colour cassette to reach the market at a reasonable price is the Sony three-quarter inch magnetic tape. The record and playback unit sell for $1,395 and the playback unit for $995. The blank one-hour colour videotape cassettes cost $35 each. The units will attach directly to the TV set, thus eliminating the need for an additional monitor. However three-quarter inch colour cameras are still quite expensive, costing from $11,000 upwards.

Another TV recording device is the Teldec Disks, being developed by AEG-Telefunken. Prototypes of this were expected by late 1972, and full production was scheduled for 1973. A severe limitation of this device is that it seems likely to be limited to 12-minute programmes in colour. Tentative price for a standard colour player is $150 and the discs themselves may be under $1. Since the Teldec Disks will be paper thin, they probably can be bound into printed material and will be quite easy to mail. Another device which shows promise is an RCA development called Selectra Vision. This uses the holographic principle and has the advantage of being quite easy to reproduce. The advantages of Selectra Vision are primarily those of mass production. The release material is an inexpensive plastic. The print is read off optically so there are few moving parts that need repair. Most important, copy wear is negligible. Taking an international view, such tapes and movie film are the only two recording media yet discussed that can be played both in the United States and in Europe without adaptors. However, it seems that commercial production of this device is still for the future.

### *Software*

The biggest problem with videotape and videocassettes is the lack of availability of prepared programmes in the fields of science and engineering, commonly known as 'software'. As indicated previously, the principal method used at present is to make tapes of actual classes as taught on campus. So far there has been relatively little done in packaging complete courses which can then be made available to working engineers who wish to update their skills. In order to partially meet this need, the Center for Advanced Engineering Study was established at Massachusetts Institute of Technology in 1964. Financed by a $5 million grant from the Alfred P. Sloan Foundation, the centre now has its own broadcasting studio and control room.

In November 1971, Dr T. Theodore G. Overbeck, internationally known professor of physical chemistry at the University of Utrecht, the Netherlands, finished videotaping at the MIT centre a fifty-five lecture series about colloid and surface chemistry. Dr Overbeck's lectures are the core of an experimental course aimed at two different audiences: the college student and the practising engineer. In schools, the course can be used to supplement the teaching programme. In industry, it serves as a self-study programme for continuing education. Three other lecture series are on calculus, probability and random processes.

The cost of preparing the Overbeck lecture series was $150,000 plus the expense of securing the speaker. Since this venture must be self-sustaining, the film or videotape is rented to industry for three months for $450 plus $130 per student; or it is leased for a year for $3,200, or for the life of the film or tape for $7,000. Copies of the study guide needed to accompany the lectures are $15 each. For universities a one-year lease is $2,250 or a life-time lease $4,500.

## THE ELECTROWRITER OR 'BLACKBOARD BY WIRE'

One of the major devices for teaching engineering classes off campus in the United States is the Electrowriter. This is a device in which signals are transmitted by two telephone lines to a remote classroom. One signal is the voice and the other is material being written at the sending classroom and reproduced in the receiving room by projection on a screen. Thus the professor is able to lecture to the distant classes and at the same time do the equivalent of writing on the blackboard. The cost of this type of equipment and the signal transmission is far less than live television.[1] The device is known as the Victor Electrowriter Remote Blackboard or the acronym 'VERB'. Among the universities which have made extensive use of this method of teaching extension courses are the University of Illinois, Auburn University, the Associated Christian Colleges of Oregon, the University of Missouri at St Louis, Oregon State University, the University of Wisconsin, Wisconsin State University, New Mexico State University, the University of New Mexico, Michigan State University and the University of Tennessee at Knoxville.

The Electrowriter teaching method offers a number of attractive advantages: (a) with suitable preparation on the part of the instructor, it provides a highly efficient means of information transfer; (b) the opportunity exists for immediate feedback—that is, a student in a remote classroom may stop the instructor at any time and question him on points that may not be clear; (c) it saves the expense and wear and tear on the instructor of regular travel to the remote class; (d) it is a relatively inexpensive means of communication. The equipment required for an Electrowriter classroom including transmitter, receiver, and accessories such as screens, microphones, speakers, and data phone installation, costs in the neighbourhood of $2,500. The major operating cost is that associated with the long-distance telephone lines. Two lines are required for each class, one for the voice transmission and one for the Electrowriter signals. However, since many universities have systems of leased telephone lines, and the classes are taught primarily in the evening when these lines are not in use, there are no appreciable incremental telephone costs chargeable to the Electrowriter classes.

Evidence exists that it is possible to teach as effectively by Electrowriter as in a live face-to-face classroom situation, particularly if the professor prepares his course carefully and provides hand-out material or transparencies which can be sent ahead by mail so that the Electrowriter need not be used for lengthy mathematical derivations. The Electrowriter then can be used for filling in steps in the development and for sketching simplified versions of the figures for emphasis.

The Electrowriter system is not without its problems, however. Some of these are: (a) the small size of the writing surface and the need to write large enough for the writing to be legible at the receiving end—this is a particular annoyance to some instructors, particularly those who rely heavily on the Electrowriter to transmit lengthy derivations, etc.; (b) the need to perform regular maintenance on the Electrowriter equipment; (c) malfunctions which arise in telephone equipment such as Dataphones, switchboard contacts, etc.; and (d) interference on the telephone lines themselves which can arise when a long-distance connexion is maintained over a one- to two-hour period.

The University of Tennessee at Knoxville began its Electrowriter programme in 1966. A peak was reached in 1968 with twelve different courses given to twenty-four total class groups, involving more than 200 industrial students in off-campus locations. However, beginning in 1969 the telephone/Electrowriter instruction was supplemented with videotapes of on-campus classes. Usually a three-credit hour course meeting for three fifty-minute periods a week can be taped (with some condensation) in two one-hour tapes which are then played back at remote locations once or twice a week. In conjunction with these playbacks, a fifteen to thirty minute telephone/Electrowriter session will be

1. One of the principal producers of this type of equipment is the Victor Comptometer Corporation, 3900 North Rockwell Street, Chicago, Illinois 60618.

conducted between the remote class and the instructor for discussion of questions, homework problems, etc. A staff member at the remote classroom will assist the instructor by passing out and collecting homework problems, administering examinations, and setting up the television and Electrowriter equipment. This combination of videotape and Electrowriter seems to be working out very well, from the viewpoint of both students and instructors.

It may be of interest to consider the cost of providing a remote classroom facility for use with the present University of Tennessee system. The following major items of equipment would be required:

Electrowriter–receiver with projector ($ 1,880); transmitter ($ 1,100).

Videotape–one or two video recorder units (E1AJ format, half-inch tape) ($ 700 each); one or two television monitors ($150-325 each); portable stands or ceiling mounts for monitors ($50 each).

Additionally, it would be necessary to install two telephone lines, one with a Speakerphone, the other with a Bell Model 601A Dataset for use with the Electrowriter. The conference room used for the class should be acoustically 'dead'; usually, a carpeted floor, draperies and an acoustic tile ceiling will create the desired effect. Thus the total cost of the facility would include $4,000 to $5,000 for equipment, the cost of acoustical preparation of the classroom (variable, dependent on the condition of the room) and a monthly rental charge of about $20 for telephone equipment.

### *Interactive lecture by Electrowriter*

There has been an interesting recent experiment at Massachusetts Institute of Technology. This is known as an 'Interactive Lecture'. First a professor is asked to record a talk on a scientific topic of both general interest and also of personal interest to him. The recording is a high-fidelity audio recording with Electrowriter sketches made by the professor as he goes along. Recorded on the second track of the tape, the Electrowriter makes it possible for the professor to illustrate what he is saying in much the same way as he would use a blackboard. The system has two units, a transmitter on which the speaker writes with a pen on a roll of paper and the receiver in which a slender arm with a stylus reproduces the writing in ink on a similar roll of paper. Students listen to the stereo tape and watch a screen as the playback unit writes out for projection those things which have been recorded by the professor to illustrate his points. At the end of the lecture the students can record their questions and later the professor makes another recording of his answers together with appropriate written material. A by-product of this system is that it has served to establish liaision between professors and students, even though they have not actually seen one another.

### *Electrowriter signals by FM radio*

A number of interesting applications of the Electrowriter system are in the development stage. One of these is the use of special FM signals to activate both voice and playback units. Thus in lieu of having direct telephone lines between the sender and receiver, a single FM station could reach all classrooms equipped with the necessary receiving devices within its normal range of operation. This might be a radius of 30 or 40 miles. The signal could be broadcast at a special frequency different from that normally employed by the station and would not interfere with the station's regular operation.

A second interesting device is one which transmits signals relayed from a satellite. Recently an experimental class was conducted by this method in which the sending unit was located at the University of Wisconsin and the receiving classroom at the Stanford University in California, a distance of some 2,000 miles. Excellent reception was obtained.

*Summary of costs*

The minimum cost of equipping a transmitting studio and a receiving classroom is as follows: transmitter \$1,100; receiver \$1,270; projector \$610; screen \$20 or \$3,000 total. This will permit one-way visual and two-way audio signals. However, if the transmitting area is to be used as a classroom, it will also require a receiver, projector, and screen; and if the remote classroom is to be equipped to ask questions by writing, it will also need a transmitter, require dual equipment, or \$6,000 total investment.

Obviously the major expense in this type of system is the use of the dual telephone lines. In the United States, use is frequently made of lines which are leased by the month. Since leasing cost and long-distance rates vary so much within different countries, no attempt will be made to evaluate this portion of the cost. Obviously if FM broadcast can be used, this may be the least expensive way of reaching many classrooms within a limited geographic area.

## AUDIO SYSTEMS

*Telenetwork*

The medical profession has made extensive use of long-distance conference telephone hook-up as a way of teaching non-credit extension classes on a regularly scheduled basis. Physicians, nurses or medical technicians may assemble in classrooms over a wide geographic area and listen to updating lectures originating at central points on the system. This is probably the least expensive method of all since only a single telephone circuit is in use. At each location, a co-ordinator projects previously prepared visual material.

Kansas State University now presents credit courses in sixteen cities throughout the state of Kansas by the Telenetwork device. The method of operation is essentially the same as that described above for the medical profession. The courses have been well received and the system is expanding. The University of Wisconsin has also had limited experience with this type of teaching.

The audio system described here lends itself very well to teaching of liberal arts subjects and general interest material. However, for the teaching of courses in engineering and mathematics, the instructor is greatly handicapped if he is not able to derive equations or in other ways make use of a blackboard. Thus it appears that the straight audio system is of limited use for the teaching of scientific material.

*Audio cassettes*

Probably no single technological development has influenced remote teaching as much as the development of the audio cassette. This has permitted the taping of lectures and their playback to students at remote locations in very inexpensive devices. A normal recording and playback unit costs about $40 and one for listening only is about $30. Thus the student can sit in the privacy of his home, in his dormitory room or at his desk at his place of employment and listen to the voice of an expert in the subject he wishes to study. It is also possible to use listening devices while driving or riding in a car in which power is obtained by plugging into the automobile's cigarette lighter.

Many professional societies are now making whole courses available on audio cassettes. These may be either purchased or leased. In addition, many of them make recordings of professional society meetings and of special lectures. These are made available to members at a very modest cost.

Textbooks using a combination of printed material and audio cassettes are beginning to be used at Oklahoma State University. We recently considered the purchase of a basics electronics laboratory manual in which a manual plus twenty-five cassettes were offered to guide students through a total of 100 exercises. The cost was rather high, with the first set of twenty-five cassettes costing $165 and each student laboratory manual $6.95. Each succeeding set of cassettes was priced at $90. Although this is rather expensive for an individual to purchase, a university can invest in a few sets of these which can then be checked out by the students and used by them for doing the laboratory work entirely by themselves. We expect in the near future to see many combination printed and audio cassette textbooks.

## NEW DEVELOPMENTS

### *Cable television*

Another recent development which is certain to affect the technology of continuing engineering education is the use of a coaxial cable for transmitting television signals directly into the homes of receivers. Originally this was known as community antenna TV or CATV. Although initially it was designed to serve small communities geographically remote from broadcast centres, it is now spreading to major cities as well as small towns. It allows the transmitting of special educational programmes originating locally directly into the home receivers. Because of the low cost of the service to the individual, which is currently about $5 per month, the concentration of receivers in any given community is extremely high. Not only can the TV set owner tune in on all of the channels which are located near him, but will be able to see programmes from a number of distant stations as well as the special educational programmes referred to above.

Cable TV is still relatively new. It has, however, had a very rapid growth and it is predicted that by 1980 from 60 to 90 per cent of all sets will be connected to cable. This market for programmes should have a strong influence on the development of specialized courses or software to meet particular educational needs.

### *Use of satellites*

An organization called the Federation of Rocky Mountain States located in Denver is currently completing a United States Department of Education planning contract for a Rocky Mountain Educational Technology Demonstration, which will utilize the National Aeronautics and Space Administrations ATS-F Satellite scheduled to be launched during 1974-75. The goal of this project is to demonstrate the effectiveness of a total instructional programme which is technologically based. It will utilize all the communications system of the region–public television stations, cable TV stations, radio frequencies, and appropriate transmission systems–in conjunction with the satellite to transmit instructional materials to children and adults in the region. The demonstration will focus on two content areas: early childhood education, and career development. If successful, this project could be the basis for future world-wide technological education, utilizing satellites.

### *Compressed speech*

The average person speaks at a rate of 110 to 160 words per minute, however, people have been trained to speak at rates as fast as 300 words per minute. This faster rate is attained by greatly reducing the lengths of pauses between words and syllables as well as between sentences. At very rapid rates, the vowel sounds are also shortened. Studies have shown that there is little or no degradation in information retention by the listener at rates in excess of 300 words per minute. The equipment has now been developed for increasing the word rate in a manner similar to a fast talker, by shortening pauses and vowel sounds. The machine adapts to different speakers so that even the slowest speaker can be recorded at rates of 300 words per minute or greater. A device for doing this is called VOCOM I.[1] With this equipment, it is possible to shorten audio tape recordings by a factor of three, thus greatly conserving time for listening to recorded material.

Compressed speech is not new. In September 1968, the *Journal of Communication* produced a special issue on this subject. This is recommended reading for those who have a special interest in this subject. Since the device is reasonably priced and relatively trouble free, it shows great promise in preparation of audio material for continuing engineering education. For example: a normal one-hour lecture on tape can now be remade into one which can be heard in only 20 minutes.

### *The picturephone*

Another technological development which shows considerable promise for continuing engineering education is the picturephone. This has been under development by Bell Telephone Company since 1927 and has had public exposure since about 1965. The first commercial installations were made in Pittsburgh, Pennsylvania, in July 1970, and in 1971 it was installed in the cities of Chicago, Springfield, Illinois, and Washington, D.C. By 1975 it will be offered in several other cities. Although no cost data are given, the units finally made available to the general public are expected to be priced competitively with other means of communication.

The unit consists of a picture display unit (containing the camera, picture tube and loudspeaker), a control unit (containing the transmitter and video control buttons), a twelve-button Touch-Tone telephone, and a service unit (containing the power supply, logic circuits, and transmission on equalizing

1. Manufactured by the PKM Corporation, 1976 Ryan Avenue West, St Paul, Minnesota 55113.

circuits). The first three units are colour-co-ordinated and intended to fit in harmoniously with any decor. The service unit is installed out of sight and can be up to 85 feet (26 metres) away from the other three.

The picture tube, which is 5 by 5.5 inches (13 x 14 centimetres), displays a picture in the same way as a home TV set. That is, it displays thirty frames a second using odd-even line interlace to give sixty fields a second, thereby yielding an acceptable 'flicker' rate. Of considerable significance is the fact that the speaker may hold in front of the 'camera' any figure which he wishes the receiver to see. Thus, it is quite possible that the picturephone can be used as another means of lecturing live to remote audiences. However, in view of the cost, it seems most probable that the device will be used to supplement other techniques such as videotape and will be used primarily in discussion sessions between students and professor after the class has viewed the lecture by videotape or by recorded electrowriter.

## SUMMARY REMARKS

A widely used technological tool in continuing engineering education is live television with talkback facility. This enables contact between the professor and the student and ensures the best learning environment. Students learn as well by this method as they do from personal instruction on campus and in many instances, TV instruction is felt to be superior. However, the cost of this instructional method is high. Usually initial investment is more than $1 million and operating costs are approximately $30 per channel per hour. To be economically feasible, student enrolment must be very high. This is the best way of reaching a large number of students in a wide geographic area.

The next most satisfactory method is the use of videotape in which either the actual classroom scene is recorded or the course is taped in a studio. This requires some type of communication between the students and the professor by telephone, by Electrowriter, or by occasional visits by the professor to the class. This is less expensive than live television but consumes a considerable amount of the professor's time. The development of video cassettes is expected to greatly increase the use of videotape for instruction.

A third device which is showing considerable promise is the use of the Electrowriter in which the voice and the equivalent of blackboard writing is sent live to distant classrooms. Like live TV, this requires the students and professors to be in a classroom at the same time although at different geographic locations. A recent modification of this technique is the recording on stereo tape of both the voice and the Electrowriter signal. This recorded message can then be sent to students at distant locations and serves essentially the same purpose as videotape. The major cost of the Electrowriter system is the rental of telephone lines. Recent experimentation on the use of FM radio for broadcasting of both voice and signals to activate Electrowriter type devices shows promise for reaching distant classrooms with voice and blackboard at moderate costs.

Transmission of voice-only to remote locations through a conference telephone hook-up is being used extensively in the medical profession. It has limited application for teaching of engineering courses, but can be used if adequate visual material is prepared in advance and arrangements are made for projecting this on the screen at the proper time during class presentations.

The most widely used and relatively inexpensive technological development is the audio cassette. Although it gives only verbal presentation, it has the advantage of low cost and ease of use. This is being used extensively for non-credit short courses, coverage of technical meetings, special lectures and as auxiliary to live classroom lectures. The development of devices for compressing speech is expected to further enhance the effectiveness of audio systems.

The picture telephone is expected to become widely used within the next five years and it shows promise as a device for off-campus instructional purposes. This is expected to make the telenetwork system effective for the teaching of engineering courses.

The development of software or prerecorded engineering courses has been very slow. As the hardware, such as video cassettes at low cost, become available, this void is expected to be filled.

For low-cost transmission of information to a very large body of receivers over a wide geographic area, the best hope is for the development of television signals or Electrowriter signals relayed from satellites. An active research and development programme is under way in the Rocky Mountain area of the United States and is expected to have a great influence on continuing engineering education in the future. This, coupled with the use of TV cable for the transmission of information directly into homes, promises exciting developments in continuing engineering education in the next decade.

*References*

1. Charles A. Martin-Vegue Jr., Albert J. Morris, Jerome M. Rosenberg and Gene E. Tallmadge: 'Technical and Economic Factors in University Instructional Television Systems', *Proceedings of the Institute of Electrical and Electrical Engineers,* Vol. 59, No. 6, June 1971, p. 946-53.
2. Paul D. Arthur and Richard S. Leavenworth: 'Project GENESYS Short Courses', *Continuing Engineering Studies Series,* No. 2, 12-14 November 1967, p. 11-22.
3. Joseph M. Pettit and Donald J. Grace: 'The Stanford Instructional Television Network', *IEEE Spectrum,* Vol. 7, No. 5, May, 1970, p. 73-80.
4. G. J. Maler, P. F. Ostwald and S. W. Maley: 'An Evaluation of Videotaped Engineering Courses at the University of Colorado', *Frontiers of Education Conference,* Tucson, Arizona, 24-27 April 1972.
5. Edward McCabe: 'The Video Cassette 1928-1971', *Training in Business and Industry,* October 1971, p. 43-60.
6. *Chemical and Engineering News,* 15 November 1971, p. 82-4.
7. *Journal of Petroleum Technology,* December 1969, p. 1559.

# Applications de la technologie à l'éducation permanente (résumé)

Professeur Monroe W. Kriegel

De nombreux et passionnants progrès techniques ont été réalisés dans le domaine de l'éducation du second degré et de l'enseignement supérieur. Parmi ces progrès figurent les suivants : emploi de la télévision, en direct ou sous forme d'enregistrements sur bandes magnétiques ; application du téléscripteur *(electrowriter)* ou "tableau noir par câble" ; large emploi de conférences enregistrées ou de systèmes d'écoute, tant au moyen de bandes magnétiques que de conférences en direct par téléphone ; mise au point de textes pour l'enseignement programmé ; utilisation de conférences par multiphone ; utilisation de câbles multiplex pour la transmission de signaux de télévision ; emploi de satellites pour la transmission de signaux de télévision et de signaux en modulation de fréquence. L'auteur étudie ces divers procédés du point de vue des méthodes utilisées, de l'efficacité de chacun comme moyen de présenter des connaissances nouvelles, et de leurs coûts respectifs.

# Reports
# of working groups

# Rapports
# des groupes de travail

# IV

# The place of continuing education

## REPORT OF WORKING GROUP 1

The group comprised: Messrs Piguet (Chairman), Carlson (Vice-Chairman), Virkkunen (Reporter), Pelegrin, El-Kholy, Goedkoop, Ihaumuotila, Krapfenbauer, Myréen, Novak, Nielsen, O'Donovan, Petrik, Rousi, Sjögren, Steichen.

The group started by challenging the name given to it. It was agreed that its task was to include a discussion of the importance of, rather than the necessity of, continuing education for engineers for the following reasons:

Technology continues to develop and to proceed along new paths.

The functions of the engineer change during his working life.

Engineers change jobs, either by choice or due to changes in employment and in manpower structure.

The individual has a legitimate and strong urge to acquire new knowledge.

International competition forces every nation to keep abreast of modern developments.

Any system of basic education cannot, by itself, cope with these urgent needs.

The planning and implementation of continuing education programmes can provide strong and mutually beneficial links between educational establishments, industry and the State.

Although there is world-wide recognition of the need for continuing education of engineers, a definition of the place of continuing education calls for a good deal of study and analysis of the existing situation.

While there is appreciable variety in the number, nature and scope of continuing-education programmes, there is, as yet, little comprehensive and systematic analysis of the needs for continuing education, present or future.

It is recommended that working groups be set up in each FEANI member country in order to: (a) develop the survey already initiated by FEANI; (b) analyse the needs for the continuing education of engineers; (c) prepare for a meeting devoted to this matter.

The group recognized the existence of certain basic issues which call for sustained attention in order to define the place continuing education should occupy in the community, namely: the need for co-ordination between the parties involved (e.g. government, industry, institutions, professional societies, regional bodies for engineering education and international organizations); the need for co-ordinated operations with existing educational systems; the definition of the responsibility for financing of continuing education; the various possibilities offered for participation; the credits and benefits.

It was noted that there are different types of continuing-education courses existing, including post-graduate credit courses, industrial training courses, and courses offered by professional associations and other bodies.

It was recommended that continuing education by its very nature and purpose should show great flexibility and allow for considerable freedom of individual choice.

# The recipients of continuing education

## REPORT OF WORKING GROUP 2

The group comprised Messrs Levy (Chairman), Niemi (Vice-Chairman), Peippo (Reporter), Hahn, Obach, Bureau, Buschmann, Hagy, Hernes, Kanttinen, Nordman, Virkkala, Zürcher, Gonzales, O'Donnell.

The working group recognized that it should deal with the recipients themselves, the people towards whom the whole effort of continuing education was directed. All the members of the group happened to come from FEANI countries so the report must be read in this context.

Who are the recipients? Clearly they include primarily the practising engineers who attend and participate in continuing-education activities. But there may also be secondary recipients, as Professor Hahn emphasized in his paper, including many sectors of society. Also the teachers themselves may in the process receive continuing education from the very students who attend from industry. Beneficial and interesting as these secondary benefits may be, it was decided that discussion should be mainly towards the chief recipients, the practising engineers who are able to profit directly from continuing-education activities in engineering.

The purposes of continuing education include the updating of knowledge in a specific field, the expansion of knowledge within a field or an associated field, and the expansion of knowledge into other fields, such as management.

Discussion was concentrated mainly toward the first purpose, the updating of knowledge in a particular field, because industrial members in particular felt that this was a most important aspect, and one in which the members of the group had the most individual experience and so felt most able to make a contribution. Other areas ought perhaps to be explored by other people in any continuation of these discussions.

By 'updating of knowledge' is meant continuing-education activities which enable an engineer to perform his present job, or something like his present job, more effectively.

To be most useful to the recipients, continuing education can take many forms and must be seen to include not just courses but also the study of engineering literature, and the use of other media of communication. There are many methods of teaching and it was deemed vital that the method used should be closely adapted to the recipient concerned at that particular time and also to the particular subject-matter. Groups using mainly discussion methods may often be appropriate, while formal lecture courses or other means may be more appropriate for groups of particular ages or backgrounds of formal education and accumulated professional experience.

It was agreed that activities coming under the heading of continuing education would not generally lead to recognized qualification such as an M.Sc. degree or a diploma. Nevertheless, it would be beneficial from the point of view of motivation and retention of what has been learnt to have some method of evaluation and assessment either during the courses or at their end.

This brought the group to perhaps the key question: how do we make continuing-education activities grow? How do we motivate? There is a large and growing market in continuing education but there are many problems involved in ensuring that the right information reaches the right people at the right time.

The group agreed that the engineering societies in each country had a central role to play in the dissemination of information and the collection of opinions of firms, of individual members, and others, and in the choice of the fields. The opinion of firms and individual members as to what they felt they needed should be canvassed, though it was accepted that this was certainly not the only factor to be taken into consideration in planning continuing education activities.

It was recognized that a considerable amount of work is already being done by means of questionnaires, announcement of courses and the like but a higher level of activity should be urged within national societies and collectively through FEANI. It was mentioned several times in the discussion that, to be useful to the prospective recipients, announcements about actual or possible courses should include not just a title but a careful statement of the ground to be covered, the depth of treatment, and the necessary prerequisites. The group was most anxious that firms and individuals should gain as full a picture as possible of each course so as to be able to form a judgement about its value to them.

The question of cost was also discussed, as a sideline to the question of motivation. In particular, some members of the working group felt that a small proportion of the cost should be borne by the recipient himself on the grounds that this would increase his involvement. Others disagreed and thought that the entire cost should be carried out by firms and government agencies. This led the group to consider the question of the provision of a statutory quota of time to be devoted by all engineers to continuing education and it came to the conclusion that this important but admittedly difficult question merited much deeper consideration by FEANI. In particular, it would be useful to collect data on the amount of money now spent within all FEANI countries on continuing education for engineers. In principle, the group would like to see progress made on the question of a certain proportion of time or money being set aside annually for this purpose but recognized that a universal figure for a percentage of time or money may not be possible owing to different circumstances between countries and between professions in any one country.

In order to convince authorities that statutory provision of time or money may be worth while, one can perhaps stress the rapid economic return which may be derived from continuing education. Unlike earlier phases of education where there is a considerable time-lag between the investment and the return, continuing education can involve the absorption of new information and ideas one month, and immediately afterwards the direct application of this new knowledge.

The group sees a continuing and growing need for management and allied courses for engineers but believes that one should not run the risk of turning engineers into bad managers and there is therefore a need for careful counselling and the provision of advice as to the best time to undertake such courses.

The working group recommends that:

1. The recipients be kept aware that continuing education in engineering can take many forms. Those concerned with teaching should ensure that the teaching methods are always closely related to the interests and backgrounds of the recipients and to the subject-matter.
2. The engineering societies play a central role in seeking views from industry, universities and individual practising engineers as to present and future needs in continuing education.
3. Information about courses and other activities include sufficient details to enable individuals to judge the probable value to themselves.
4. Further consideration be given by FEANI to the question of urging the provision of a statutory proportion of each engineer's time, or the provision of an annual sum of money, for purposes of continuing education.
5. Advice be available to individuals on the timeliness for them of particular types of courses, especially with regard to management and similar courses. This advice might, for example, be given by making available the accumulated experience of others.

# Planning of continuing education

## REPORT OF WORKING GROUP 3A

Working Group 3A comprised Messrs Profos (Chairman), Tymowski (Vice-Chairman), Porra (Reporter), Jerez Juan, Ando, Debruyin, El-Erian, Keskinen, Kings, Reistad, Rindt, Tassios, Tunkelo, Vervoort.

With regard to the responsibility to take the initiative in organizing continuing-education courses, the group felt that the present situation was not wholly satisfactory, and agreed unanimously that somebody representing engineering societies, universities, and eventually governments should be responsible for taking initiatives. It was suggested that in many countries the engineering societies should play the leading role in this body.

It was thought that information about training needs and goals could be determined by experts, by sample surveys, and by individual questionnaires to all prospective recipients. Interested industries should also be consulted and national development plans used as sources of information. As the organization and evaluation of questionnaires need special knowledge, it was the opinion of the working group that specialists in the conduct of such surveys should be used. Similar careful work is needed in collecting and evaluating profiles of potential participants.

The development of course curricula was stated as often being done by a small group that fixes the subject and the framework of the course, with the chosen teacher working out details. This was seen as a satisfactory procedure.

Opinions were divided with regard to the choice of teachers, as to whether preference should be given to those from academic circles or from industry. In some countries handbooks and other materials or teaching methods are available for inexperienced teachers. The working group felt that the lecture should normally be supplemented by prepared notes.

With regard to financing, the general opinion of the working group was that continuing education is of major national interest and therefore governments should give financial support, but that industry and normally participants could also contribute.

It was agreed that feedback and past experience should be used as an element in planning future courses, but that if questionnaires are used, very careful evaluation is required.

The working group recommends the following:

1. That when gathering information about training needs and goals the interested industries also should be consulted. Furthermore the national development policies should be used as a source of information.
2. As the organization and evaluation of questionnaires on training needs and goals need special knowledge, specialists in the conduct of such surveys should be used.
3. Concerning the choice of teachers, it should be a guideline that scientific fundamentals are a necessary component of every course.
4. In some countries, handbooks and other material on teaching methods are available for inexperienced teachers. This approach should be expanded.
5. Prepared notes should be furnished to supplement continuing education lectures.
6. Continuing education is of major national and social interest and, therefore, the government should give financial support.
7. Experience gained on completed courses should be used regularly as a source of information for future planning.

## REPORT OF WORKING GROUP 3B

Working Group 3B comprised Messrs Titchener (Chairman), Rose (Vice-Chairman), Saarialho (Reporter), Gestrelius, Biedenbach, Ferrari-Toniolo, Gysin, Farvis, Knop, Laasonen, Logothetis, Peters, Szebenyi, Salmela.

With regard to the planning of continuing education for engineers, the working group was unanimous that it was not possible to do this on a purely national basis, even though some co-ordination is highly desirable, but that the organization should normally be decentralized, in a way that would vary according to the particular circumstances in each country.

The working group saw the problem of who should do the planning existing at four main levels, namely: the requirements of the individual; the role of industry and the educational institutions; the role of the professional societies; the role at the national level of guideline planning.

At the individual level there must be an urge to advance in life through self-study and participation in continuing education. Each individual must work out a career profile and try to work to this aim through various courses and by all other available means. This is of utmost importance for any planning and effective implementation of continuous education.

The educational institutes and industry are on purpose considered at the same level, to emphasize the necessity of good liaison and co-operation between them. In the planning of continuing education they must work very closely together if successful continuing education is to be achieved.

The professional societies must play a more important role in continuing education in the future, compared to that they have played so far. They could help in the planning of continuing education by carrying out job analyses and finding out where the existing programmes of continuing education are insufficient to meet the real needs of the profession. In this respect they could help both industry and the educational institutions to come together and to evolve worthwhile programmes of continuing education.

At the national level, the group felt that there is a need to have a sort of central body or agency which should in some way co-ordinate the activities carried out at various levels and through various bodies in the field of continuing education.

In some countries this agency could conceivably stimulate the preparation of necessary legislation designed to provide stimulation and support for continuing-education programmes.

The working group noted, with regard to the preparation of curricula, the studies being carried out by VDI (Ingenieur Von Morgen Hilmen, Peters & Polke, VDI Model X-172), other work such as that described in the paper by Mr Gestrelius, and the work being undertaken by the Committee on Education and Training of the WFEO.

It was pointed out that it was necessary to distinguish two goals in planning curricula. The first is the improvement in the qualification profile of the individual, and this is to the benefit of the whole nation, while the second goal of improvement in job qualifications gives benefits primarily to industry, often immediately.

The group agreed that job-analysis methods were applicable to planning curricula for continuing education, and that it would be possible to apply these methods to internal training programmes carried out by large employers, as well as to work carried out in some universities. It was mentioned that these techniques were already being used in some universities, notably in the United States.

In the discussion of the problem of whether basic and continuing education could be integrated to form a structurally coherent whole, the question of the extent to which continuing education should be made compulsory also arose, but this somewhat controversial question remained unresolved.

The group agreed that it saw continuing education as a part of the whole education process, but that future experience will reveal the appropriate mechanisms for such integration.

The view was expressed that one might expect in some countries a shortening of the period for basic education in the future. This being so, the role of continuing education would become increasingly important for engineers planning their future careers. If the basic-educational programmes are to be shortened, it is important to ensure that the subjects taught in them are not subject to rapid obsolescence. After a student has completed his basic-educational programme, he could expect to go on to more specialized areas of study suited to his particular career intentions. The group saw the value of a series of staged qualifications based on incentives for the individual's advancement in his professional career. The group noted that the major portion of educational expenditure is at present used for the education of persons under the age of about 25 years. It foresees the necessity of altering the balance of expenditure so that a much larger proportion will be devoted to persons who are engaged in extending their education in the more mature years of life.

The group considered it essential to fund continuing education soundly and adequately. It felt that the body funding basic education should in general be the body funding continuing education as well. The general view was that the financial burden on the individual undertaking continuing education should not be too great, because he must make other sacrifices as well in order to participate in continuing education programmes. In many cases industry is the direct beneficiary of these programmes; the group thus would expect that industry would carry some of the financial burden. In some countries where continuing-education programmes have been working successfully for some time, and particularly in large countries where the organization of education is complex, a single national means of funding continuing education is probably not possible at this time.

The group saw the need to organize continuing-education programmes both in educational institutions and in the workplaces of the recipients, depending on local circumstances.

As regards the possibility of educational institutions coping with the very rapid increases likely in the scale of continuing education, the group agreed that the first need was for separate and adequate budgets. It was thought necessary to set up a separate department or unit within the educational institution to handle the organization and operation of the continuing-education programmes, and that close liaison between industry, the educational institutions and professional bodies was required in such undertakings.

The group recommended:

1. That the possibility be explored of obtaining funds from foundations (such as the Kell Foundation) or other sources, for setting up centres of continuing education for engineers in Europe.
2. That the existing Education and Training Committee of FEANI should take responsibility for keeping under continuous review the needs for continuing education of engineers.
3. That, at the national level, there be set up a central agency for the purpose of co-ordinating the continuing-education activities carried out at various levels and through various bodies in the field of continuous education in the country of concern. This agency could lay down broad guidelines and, wherever appropriate, help in preparing legislation whereby continuing-education programmes are supported.
4. That, in planning programmes of continuing education, note be taken of the possibility of using job-analysis and related methods for designing curricula, and attention be drawn to existing work in progress in this field.
5. That means be pursued of increasing the financial support for continuing education. This will call for an increase in the total support for education, but may also involve alterations in the balance of educational expenditure as between basic and continuing education.

# The means of continuing education

## REPORT OF WORKING GROUP 4

The group comprised: Messrs Ham (Chairman), Leonard (Vice-Chairman), Mrs Blomstedt (Reporter), Miss Törnudd, Messrs Corbett, Kriegel, Mizushina, Bucovics, O'Donnell, Hohenthal, Kinnunen, Pessoa, Sorel, Yadarola, Sherehan.

The working group recognized several distinct goals of continuing education, such as updating, re-education, and further education. Each application is characterized by theoretical and/or practical needs for learning which call for the use of the aural, visual and physical senses (including laboratory experiments).

Whereas re-education may most appropriately be carried out on a full-time basis, up-dating and further education may be achieved by diverse means.

The need for studies of the process of learning in continuing education was emphasized. Such work should elicit the support of educationists, psychologists and engineering practitioners, and should concern itself with the motivation of the participants and the effectiveness of their experience.

The group believes that continuing education is now of such significance in the development of all countries that its effectiveness can only be assured by the provision of appropriate financial support through such mechanisms as taxation, fringe benefits, or rights in the contract of employment. There is a need for industry, professional practitioners and governments jointly to manifest their commitment in this matter.

It is essential for governments and enterprises to understand that learning is a part of working. The allocation of time to learning while working must be determined in the light of many different circumstances, and no absolute standard could be envisaged. Differential privileges for continuing education may be granted to engineers performing different roles such as design and development. Engineers should be encouraged not only to develop motivation for self-learning but also to act as teachers among their immediate colleagues.

With respect to the technologies employed for instruction, the group affirms that these must be seen as means and not as ends. Except in the field of computer-based information services, there appears to be no absolute need for any specific technology in continuing education. If there are no financial limitations, the choice of technology will be determined by considerations relating to the scope and detail of the application. While various technological means exist to disseminate information, there remains a great shortage of well-prepared programmes. The generation of programme content will remain an important issue.

The group considers it essential to pay great attention to the assessment of the cost and benefits of the use of technologies, and to admit failures honestly. The assistance of educationists, psychologists, practitioners and recipients will be necessary. The assessment of the various technologies is as yet in a primitive state.

The desirability of each country evaluating the means of continuing education by assessing what is feasible and relevant within its own system and plans was affirmed.

To assist member nations in reaching decisions on means, the group recommends strongly that one or more films or similar devices be produced and regularly updated to illustrate available information services, methods for using technologies, and examples of actual forms of continuing education whether technologically intensive or not. These conference films should be made widely available. A primary purpose of such information on the state of the art should be the avoidance of duplication of national efforts. Enterprises should if possible designate a person to act as interpreter of the state of methods relevant to its operations. FEANI may wish to consider the desirability of setting up a working group to investigate this area of work, in collaboration with Unesco and WFEO.

The working group thus recommends that:

1. Processes of learning in continuing education be explicitly studied with the assistance of educationists, psychologists and engineering practitionors.
2. Practical support for continuing education be assured by suitable national processes of taxation and the provision of industrial benefits in the conditions of employment.
3. (a) All agencies that utilize the services of engineers be urged to recognize learning as a part of the responsibilities of professional work.
   (b) Differential privileges with regard to continuing education be granted to engineers performing different roles.
   (c) Engineers within enterprises be encouraged to assume teaching responsibilities for their colleagues.
4. (a) New technologies of instruction be clearly understood to be intermediary means and not ends.
   (b) The decision to use new technologies be based upon careful analysis of their feasibility in the local context (financial, availability of programme materials) and accumulated assessments or their effectiveness.
   (c) That the assessment of the usefulness of new technologies be considered an essential part of the study of the processes of learning in continuing education.
5. A reference film (or other suitable device) be made widely available to illustrate computer-based information services, methods of using instructional technologies, and examples of actual practices in continuing education whether they are technologically intensive or not.
6. Enterprises employing engineers designate a person to act as an interpreter of the state of methods relevant to their operations in continuing education.
7. FEANI set up a working group on continuing education within its Education Committee to co-operate with Unesco and WFEO.

# International co-operation in continuing education

## REPORT OF WORKING GROUP 5

The group comprised: Messrs Barry (Chairman), Broida (Vice-Chairman), Nurmi (Reporter), Bertrand, Clogenson, Evstafiev, Irgens, Janssen, Klus, Mrs Monticone, Messrs Mocanou, Sharp, Slungaard, Vermehren, Vlugter, Warner, Yackovlev.

The working group reported that the continuing education of engineers has become widely accepted in many countries and will almost certainly increase in importance in the years ahead. Therefore the planning for continuous education, including the basic studies, design of physical facilities and curricula should take this into account. Furthermore, all engineering students should be made aware that their basic engineering studies are only the beginning of their learning.

The group recognized that continuing education for engineers is being developed most rapidly in the major industrialized countries, but that developing countries are often in even greater need for such programmes. This is partly because the engineers in these countries often acquire greater responsibilities at younger ages than their colleagues in industrialized countries, and because they have less opportunities for self-education or on-the-job learning. International co-operation amongst engineers was seen as vital for the solution of these problems.

The recommendations of the working group are as follows:

1. The group recommends to Unesco that it establish a working group to advise it in the formulation and implementation of a programme of continuing education of engineering personnel. This group could include engineers, technician engineers and engineering technicians.
2. A series of meetings should be arranged for exchange of experience, demonstration of techniques and future planning of continuing education of engineering personnel. Such meetings should be held at regular intervals, and the results widely disseminated. The location and sponsoring organizations may vary from meeting to meeting, taking into account the interests of intergovernmental, non-governmental, regional and national organizations in the problems involved.
3. As examples of implementation of the above recommendation, FEANI is encouraged to recommend to OECD and the Council of Europe that they participate in such action in the field of the continuing education of engineers.

4. As a further example of the above, the working group supports the suggestion that within regions the directors of continuing education centres and programmes should establish forums and periodically meet to exchange experience. In some instances it may be valuable to hold meetings of those responsible for programmes in specific fields, for example in chemical engineering or electrical engineering. It is suggested that UATI, as a federation of international societies in specific engineering disciplines that has special status with Unesco, may be able to co-ordinate this type of activity. Furthermore, it is suggested that one or more members of the Unesco working group could be drawn from such forums and close links established between them.
5. The participation of personnel from other countries as teachers or students in national programmes of continuing education should be encouraged. Furthermore, the Unesco working group should explore the feasibility and the eventual location of national centres suitable for continuing-education programmes with regular and wide international participation. The possibility of organizing such continuing-education programmes for engineers primarily engaged in research or management should not be overlooked.
6. In countries where an organization or group of organizations responsible for the continuing education of engineers does not yet exist, the creation of such a body is recommended. In many countries the same body may also be concerned with basic engineering education, and may be widely representative of industry, education, government and engineering societies.
7. The group recommends that national and/or regional organizations publish periodically material on continuing education and that the Unesco working group promote the publication of such material.
8. The group recommends that engineering groups in the various countries be prepared to actively support the use of communications satellites in continuing education, and particularly for international co-operation in continuing education for engineers. One example of the urgency to take action may be quoted here. It refers to the present provisional availability of the 2.5 gigahertz band for long-distance educational purposes via satellite: this availability is revisable in ten years and is likely to be cancelled unless significant action is taken to demonstrate its value.
9. It is recommended that those engaged in the elaboration and implementation of continuing education programmes in advanced countries should at all times keep in mind the possibility of sharing their experience and even their educational materials with those in other countries, notably their fellow engineers in the developing countries.

# Discussion in plenary session on 24 August 1972

The discussion of the reports of the six working groups was limited by the time available. Only very few reservations or differences of opinion were expressed, and the reports received wide support from the participants.

There was unanimity on the need for increased provision for the continuing education of engineers, on the need for co-ordination on a national scale and consultation on an international scale, and on the responsibility of the individual to take advantage of opportunities for his continuing education in pursuing his planned career.

One speaker thought it unwise to predict future trends in engineering education, and in particular questioned the assumption that the duration of basic education for engineers would become shorter. He also wished to stress that, as well as industry, society as a whole stands to benefit from the continuing education of engineers.

There was some discussion on the categories of personnel covered by the recommendations. While most participants appeared to favour their application to the whole spectrum of engineering manpower, comprising engineers, technician engineers and technicians, there were several who maintained that they should be specifically restricted to the type of personnel that are within the member organizations of FEANI.

Another participant stressed that in his view it was essential to distinguish a truly professional engineer from what may be loosely described as a qualified engineer, saying that the education and training and continuing education of a professional engineer must include appropriate emphasis on his responsibilities towards the public in the discharge of his employment. Not only do these comprise public health and safety, but they also stem from his role in the community in respect of the human, social and environmental aspects of his work. These broader aspects are likely to become an increasing constraint in the future, and continuing education must have such matters within its scope.

A number of speakers stressed the need to organize meetings to exchange experience, suggesting that regional bodies such as FEANI should be active in collaboration with Unesco. Also, it was thought important to bear in mind the needs of the developing countries, and that any major engineering project in these countries should include a training component for local engineers.

These who contributed to the discussion on 24 August 1972 included Messrs Profos, Titchener, Tassios, Leonard, Piguet, Krapfenbauer, Barry, Broida, Tymowski, El Kholy, Levy, Ham, Bose, Sharp, O'Donnell, Gestrelius, Vlugter, Niemi, Kriegel.

# Conclusions and recommendations

# Conclusions et recommandations

# Beschlüsse und Empfehlungen

# V

# Conclusions and recommendations

This seminar, organized by FEANI and Unesco, has provided the opportunity for delegates and participants from many countries to exchange views on continuing education for engineers.

They have concluded that the needs are now clear in all countries for the engineer to receive fresh information and understanding throughout the whole of his working life. The first need of all is based on the role of the engineer in society to improve the conditions in which his fellow citizens live, by meeting the increasing demand for food, housing, communications, and all other services. The improvement is bound up in applying the sciences and technologies so that they continue to provide for the needs of society for new products and for services, but do so in more economic and rational ways. The advances in technology require intellectual understanding and appreciation for which initial education and training is no longer sufficient. There is a need for the continual updating of the knowledge of the engineer and for the maintenance of his competence in those intellectual skills that are not in daily use but are vital for the practice of his profession.

In order to meet his obligations to perform efficiently and safely, the engineer needs to keep himself informed of technical developements. In addition, he must be able to face the need to change his job at some time, either because of developments in his own working situation or because there is no longer a demand for his special abilities. In either case, he may need to deepen his knowledge of the skills he is using or to acquire new ones, particularly where the development of his career calls for other skills in the field of management.

The recommendations of this seminar have taken into account these needs and the ways in which institutions can contribute to their fulfilment. The institutions concerned are society in general through government organizations, industry, the universities and higher technical schools, and engineering societies. Depending on the formal state of relations between the bodies, some method of mutual consultation and support is required so that effective education can be provided and financed.

The motive force in producing the developments which have taken place is the resolve of the individual engineer to undertake a process of re-education and in this the engineering society has often provided the vehicle. The needs now identified are so great that further progress requires the purposeful co-operation of all institutions concerned. They also require that experience be exchanged at international level and that the work of stimulation and co-ordination be put on a continuing basis.

The seminar therefore addresses a number of recommendations to the bodies which sponsored it in Helsinki during August 1972, with the request that they should take responsibility, as far as they are able, for seeing that they are carried out.

The seminar specifically recommends that:

## INTERNATIONAL CO-OPERATION

1. Existing mechanisms for international co-operation in the continuing education of engineers be strengthened, and in particular that an international working group be established by Unesco in collaboration with other interested international and regional bodies to advise Unesco on the formulation and implementation of international programmes in the continuing education of engineers. This working group should take into account the needs of all those concerned, among which there would be the interested international and regional organizations, governments, educational institutions, industries and professional bodies.

2. One role of the Unesco working group be to stimulate and co-ordinate the holding of a series of international and regional meetings and studies dealing with the problems of continuing education of engineers. A specific need is for the holding of meetings of persons directly responsible for the conduct of continuing education centres and programmes.

3. Through the above mechanism and by all other means there be free exchange of information concerned with continuing education of engineers, so that experience will be shared on an international scale. Furthermore, the sharing of teaching material should be facilitated, and consideration given to using new technologies such as satellite transmission for this purpose. Action should especially be taken to meet the growing needs of the developing countries for information and materials concerned with engineering education.

4. International participation as teachers and students in programmes of continuing education of engineers be encouraged and facilitated on both a regional and international basis, and that Unesco, WFEO, FEANI, OECD, CEE, the Council of Europe, and other bodies act effectively in this area.

5. In each country an appropriate national body able to communicate effectively with all those engaged in continuing education be created or strengthened, keeping in mind that one of the pre-requisities for effective regional and international co-operation is the existence of such an appropriate national focal point.

## THE BENEFICIARIES OF CONTINUING EDUCATION

6. It be recognized that the principal beneficiary of the continuing education of engineers is society as a whole. The individual engineer, his employer, and his profession will also benefit to considerable but varying degrees.

7. Through means such as professional societies and the employers, engineers be made fully aware of the possibilities of continuing education available to them, and of their rights in this regard. Similarly, the obligations of industry and

employers and government to provide support for continuing education should be promoted.

8. The engineering societies and similar bodies play a positive role in gathering and disseminating information on the needs for continuing education of engineers through contact with employers and industry, inaividual engineers, and educational institutions.

9. Those concerned with teaching should ensure that content and methods be governed by the needs, interests and background of the recipient engineers, and that they should themselves have the necessary background and contacts to perform this function effectively.

## MEANS OF CONTINUING EDUCATION

10. Systematic studies be made of the processes of learning that are involved in continuing education, and that this work be carried out with the aid of those working in disciplines such as psychology and pedagogy. Furthermore, it is recommended that bodies such as Unesco and FEANI promote the establishment of links between those working in this field in different countries.

11. There be effective exchange of experience on the methods of instruction and information systems available for continuing education, especially on new methodology, and that the successes and failures of various approaches be widely publicized by the appropriate professional bodies.

12. The national bodies concerned with engineering establish appropriate means so that all those engaged in continuing education of engineers, including those in industry and in universities, are made aware of the benefits and limitations of the methods available for continuing education.

13. Adequate financial means be available for the continuing education of engineers through appropriate legislation and fiscal action, and that the responsibility of the employer be clearly defined in this respect.

# Conclusions et recommandations

Ce séminaire organisé par la Fédération européenne d'associations nationales d'ingénieurs (FEANI) et l'Unesco, a permis à des spécialistes de nombreux pays d'échanger des idées sur l'éducation permanente des ingénieurs.

Les participants ont conclu à la nécessité manifeste, dans tous les pays, de faire en sorte que l'ingénieur puisse se recycler tout au long de sa vie active. Cette nécessité tient en premier lieu au rôle même de l'ingénieur dans la société, qui est d'améliorer les conditions de vie de ses concitoyens en répondant à leur demande croissante de produits alimentaires, de logements, de moyens de communication et de tous les autres services. Cette amélioration est liée à une application des sciences et des techniques telle qu'elle réponde constamment aux besoins de la société en produits nouveaux et en services, mais en le faisant d'une façon plus économique et plus rationnelle. Les progrès de la technique exigent aujourd'hui de l'ingénieur une capacité d'assimilation et de compréhension auxquelles ses études et sa formation ne l'ont pas suffisamment préparé. Il lui faut actualiser sans cesse ses connaissances et maintenir sa compétence dans des disciplines qu'il n'utilise pas tous les jours mais qui sont essentielles à l'exercice de sa profession.

Pour s'acquitter de ses obligations et le faire avec toute la compétence et l'efficacité voulues, l'ingénieur doit se tenir au courant de l'évolution des techniques. Il doit en outre être prêt à changer d'emploi soit à cause de changements intervenus dans la situation de sa profession, soit parce que ses compétences particulières ne trouvent plus à s'employer. Dans un cas comme dans l'autre, il peut lui être nécessaire d'approfondir sa connaissance des techniques qu'il emploie ou d'en acquérir de nouvelles, surtout lorsque sa carrière exige d'autres compétences, par exemple dans le domaine de la gestion.

Les recommandations de ce séminaire tiennent compte de ces nécessités et de la façon dont les institutions peuvent aider à y répondre. Ces institutions sont la société en général par le truchement des organismes d'Etat, l'industrie, l'université, les grandes écoles et les associations d'ingénieurs. Selon l'état des relations existant entre ces divers organismes, il faudra trouver une méthode de consultation et d'appui mutuel qui permette d'organiser et de financer cette formation et d'en assurer l'efficacité. Le moteur de l'évolution qui s'est produite réside dans la résolution de chaque ingénieur de subir un recyclage et l'association d'ingénieurs lui en a souvent donné le moyen. Mais les besoins sont aujourd'hui si grands que, pour progresser davantage, il ne sera pas trop de la coopération de toutes les institutions intéressées. Il faudra aussi organiser l'échange de données d'expérience à l'échelle internationale et, faire en sorte que les activités d'encouragement et de coordination prennent un caractère permanent. C'est dans cet esprit que les participants au séminaire adressent un certain nombre de recommandations aux organismes qui ont patronné le séminaire à Helsinki en août 1972, en leur demandant de veiller eux-mêmes, dans toute la mesure du possible, à leur application.

Les participants recommandent plus particulièrement ce qui suit :

## COOPÉRATION INTERNATIONALE

1. Les mécanismes de coopération internationale existants pour l'éducation permanente des ingénieurs devraient être renforcés, et, en particulier, l'Unesco devrait, en collaboration avec d'autres organes internationaux et régionaux, créer un groupe de travail international, pour la conseiller en ce qui concerne l'élaboration et l'application de programmes internationaux d'éducation permanente des ingénieurs. Ce groupe de travail devrait tenir compte des besoins de toutes les parties au nombre desquelles figurent les organisations internationales et régionales, les Etats, les établissements d'enseignement, les industries et les organismes professionnels intéressés.

2. Le groupe de travail de l'Unesco aurait notamment pour rôle d'encourager et de coordonner l'organisation d'une série de réunions et d'études internationales et régionales sur l'éducation permanente des ingénieurs. Il faudrait plus précisément organiser des réunions entre ceux qui sont directement responsables des centres et programmes d'éducation permanente.

3. Grâce au mécanisme ci-dessus et par tous autres moyens serait assuré le libre-échange d'information sur l'éducation permanente des ingénieurs de manière à garantir la diffusion des données d'expérience sur le plan international. En outre, il conviendrait de faciliter le partage de la documentation et envisager d'utiliser à cette fin des techniques nouvelles telles que les communications par satellite. Il faudrait s'attacher tout spécialement à répondre aux besoins croissants des pays en voie de développement en information et documentation portant sur l'éducation des ingénieurs.

4. La participation de professeurs et d'étudiants de tous les pays aux programmes d'éducation permanente des ingénieurs devrait être encouragée et facilitée à la fois sur le plan régional et international, et l'Unesco, la FMOI, la FEANI, l'OCDE, la CEE, le Conseil de l'Europe et d'autres organismes devraient s'y employer activement.

5. Dans chaque pays devrait être créé ou renforcé un organisme national apte à communiquer efficacement avec tous ceux qui s'occupent d'éducation permanente, étant donné que l'existence d'un point de ralliement national est l'un des préalables à une coopération régionale et internationale efficace.

## LES BÉNÉFICIAIRES DE L'ÉDUCATION PERMANENTE

6. Il est reconnu que le principal bénéficiaire de l'éducation permanente des ingénieurs est la société dans son ensemble. L'individu lui-même, son employeur et sa profession en bénéficieront aussi beaucoup, mais à des degrés divers.

7. Par l'intermédiaire de leurs associations professionnelles ou de leurs employeurs, les igénieurs devraient être mis entièrement au courant des

possibilités d'éducation permanente qui leur sont offertes et de leurs droits à cet égard. En outre, il conviendrait d'accroître les moyens nécessaires à l'éducation permanente des ingénieurs.

8. Les associations d'ingénieurs et organismes analogues devraient contribuer efficacement à recueillir et à diffuser des renseignements sur les besoins d'éducation permanente des ingénieurs, par des contacts avec les employeurs et l'industrie, les ingénieurs eux-mêmes et les établissements d'enseignement.

9. Les responsables de l'enseignement devraient veiller à ce que le contenu et les méthodes d'enseignement répondent bien aux besoins, aux intérêts et à la formation des ingénieurs, et que ceux-ci aient eux-mêmes la formation et les relations nécessaires pour s'acquitter de leur mission.

## MOYENS DE L'ÉDUCATION PERMANENTE

10. Des études systématiques devraient être faites sur les processus d'acquisition des connaissances dans l'éducation permanente, avec l'aide de spécialistes en psychologie ou en pédagogie. Il est recommandé, en outre, que des organismes comme l'Unesco et la FEANI encouragent l'établissement de liens entre ceux qui s'occupent de ces questions dans les divers pays.

11. Un échange efficace de données d'expérience devrait être organisé sur les méthodes pédagogiques et les systèmes d'information disponibles pour l'éducation permanente des ingénieurs, notamment sur les méthodes nouvelles. Les succès et les échecs des diverses méthodes devraient être largement portés à la connaissance du public par les organes professionnels compétents.

12.Les organismes nationaux qui s'occupent d'ingéniorie devraient créer des moyens permettant à tous ceux qui sont employés dans l'éducation permanente des ingénieurs, aussi bien dans l'industrie qu'à l'université, d'être pleinement informés des avantages et des limites des méthodes de l'éducation permanente.

13. Des crédits suffisants devraient être ouverts pour l'éducation permanente des ingénieurs grâce à une législation et à une action fiscale appropriées, et la responsabilité de l'employeur à cet égard devrait être clairement définie.

# Beschlüsse und Empfehlungen

Dieses gemeinsam von der FEANI und der Unesco veranstalte Seminar gab den Delegierten und Teilnehmern vieler Länder Gelegenheit, Meinungen im Zusammenhang mit der Fortbildung der Ingenieure auszutauschen.

Es wurde festgestellt, dass die Bedürfnisse für Ingenieure aller Länder gleich sind, nämlich während der ganzen Zeit ihrer beruflichen Laufbahn auf den neuesten Stand gebrachte Informationen und Verständnis zu erhalten. Die wichtigste Aufgabe bezieht sich auf die Rolle, die der Ingenieur in der Gesellschaft spielt, nämlich die Lebensbedingungen seiner Mitbürger zu verbessern, indem er den steigenden Anforderungen auf dem Sektor der Nahrung, des Wohnens, der Kommunikation, sowie jeder anderen Dienstleistung gerecht wird. Diese Verbesserung geht Hand in Hand mit der Anwendung der Wissenschaften und Technologien; es soll daher angestrebt werden, dass diese weiterhin für die Bedürfnisse der Gesellschaft neue Produkte und Dienstleistungen versehen, aber in einer wirtschaftlicheren und rationelleren Form als bisher. Die in der Technologie gemachten Fortschritte setzn Verständnis und Urteilskraft voraus für die die ursprüngliche Erziehung und Lehre nicht mehr ausreichend sind. Es ist heute absolut notwendig, dem Ingenieur ein auf den neuesten Stand gebrachtes Wissen zu vermitteln, damit dieser seine rationellen Kenntnisse, die wohl nicht täglich angewandt werden, aber wesentlich für die Ausübung seines Berufes sind, auf dem Laufenden halten kann.

Damit der Ingenieur seine Pflichten wirksam und sicher erfüllen kann, muss er sich in den technischen Neuheiten immer auf dem Laufenden halten. Ausserdem muss er dazu fähig sein, die Notwendigkeit zu sehen, zu irgendeinem Zeitpunkt seine Stellung eventuell zu wechseln, sei es auf Gründen der Entwicklung seiner eigenen Position oder weil in seinem besonderen Fach keine Nachfrage mehr besteht. In beiden Fällen ist er vielleicht gezwungen seine bisher angewandten Kenntnisse zu vertiefen oder neue dazu zu gewinnen, besonders dort, wo die Entwicklung seiner Karriere andere Fertigkeiten verlangen.

Die Empfehlungen dieses Seminars haben diesen Forderungen Rechnung getragen und Wege gefunden, wie verschiedene Einrichtungen zur Erfüllung dieser Forderungen beitragen können. Die betreffenden Einrichtungen sind ver allem die Gesellschaft im allgemeinen, vertreten durch staatliche Organisationen, Industrie, Universitäten und technische Hochschulen sowie die Ingenieurvereine. Abhängig von der formalen Beschaffenheit der Verbindung zwischen diesen Organismen, ist eine Art von gegenseitiger Beratung und Unterstützung erwünscht, so dass eine effektive Ausbildung geschaffen und finanziert werden kann. Die Treibkraft zur Hervorrufung dieser Entwicklungen soll den einzelnen Ingenieur dazu veranlassen, sich einem Umschulungsprozess zu unterziehen : für diese Aufgabe spielten die Ingenieurvereine oft eine Vermittlerrolle. Die ANforderungen, die hier nun erkannt wurden, sind so bedeutend, dass jeder weitere Schritt eine zielgerichtete Zusammenarbeit von allen betroffenen Einrichtungen erfordert. Ausserdem ist es notwendig, die Erfahrungen auf internationalem Niveau auszutauschen und die Arbeit des Antriebes und der übereinstimmung auf einer kontinuierlichen Ebene aufzustellen.

Das Seminar wird daher eine Anzahl von Empfehlungen an die Organisationen richten, die ihm in Helsinki im August 1972 ihre Unterstützung gewährt haben, mit der Bitte, dass diese die Verantwortung – soweit dies möglich ist – zur Nachprüfung, ob diese Empfehlungen auch wirklich durchgeführt werden, übernehmen sollen.

Das Seminar macht folgende Empfehlungen :

## INTERNATIONALE ZUSAMMENARBEIT

1. Die vorhandenen Triebkräfte in der internationalen Zusammenarbeit für die Weiterbildung der Ingenieure sollen verstärkt werden. Die Unesco errichtet eine internationale Arbeitsgruppe, unter der Mitwirkung anderer daran interessierter internationaler und lokaler Organisationen, um die Unesco über die Formulierung und Ausführung der internationalen Lehrpläne zur Fortbildung des Ingenieurs zu unterrichten. Diese Arbeitsgruppe soll die Anforderungen der daran interessierten internationalen und lokalen Organisationen, Regierungen, Erziehungseinrichtungen, Industrien und beruflichen Organisationen berücksichtigen.

2. Eine der Aufgaben dieser Unesco-Arbeitsgruppe wird die Anregung und Abstimmung für verschiedene geplante internationale und lokale Treffen und Seminare sein, die die Weiterbildung des Ingenieurs behandeln sollen. Es ist unerlässlich, dass diese Zusammenkünfte vor allem jene Personen ansprechen, die unmittelbar für die Leitung der Fortbildungszentren und der Lehrpläne verantwortlich sind.

3. Durch die o.a. Einrichtung und durch andere Mittel und Wege wird ein freier Austausch von Informationen über die Fortbildung von Ingenieuren stattfinden, damit so Erfahrungen auf internationaler Ebene ausgetauscht werden können. Ausserdem soll das Verteilen von Lehrmaterial erleichtert werden und die Möglichkeit neue Technologien anzuwenden, wie z.B. Vermittlung durch Satelliten, ins Auge gefasst werden.

Es sollen vor allem Massnahmen getroffen werden, um die steigenden Bedürfnisse, der Entwicklungsländer an Informationem und Lehrmaterial in Bezug auf die Ingenieurausbildung zu befriedigen.

4. Die internationale Zusammenarbeit von Professoren und Studenten zur Aufstellung von Lehrplänen für die Weiterbildung des Ingenieurs soll auf lokaler, wie auch auf internationaler Basis gefördert und erleichtert werden und die Unesco, WFEO, OECD, CEE, Europarat und EWG sowie andere Organisationen sollen ihre aktive Teilnahme auf diesem Sektor zusichern.

5. In jedem Land soll eine entsprechende nationale Organisation gegründet oder verstärkt werden, die zum Ziel hat, mit all denen, die sich mit Weiterbildung befassen, in Verbindung zu stehen, und die sich auch bewusst sein soll, dass das

Vorhandensein eines solchen entsprechenden nationalen Brennpunktes eine der Vorbedingungen für jede wirksame lokale und internationale Zusammenarbeit ist.

## DIE NUTZNIESSER DIESER WEITERBILDUNG

6. Es wurde festgestellt, dass der Hauptnutzniesser dieser Weiterbildung für Ingenieure die Gesellschaft als Ganzes ist. Das Individuum selbst, sein Arbeitgeber und sein Beruf werden ihren Vorteil in einem unterschiedlichen Ausmasse haben.

7. Durch die Berufsvereinigungen und ihre Arbeitgeber werden die Ingenieure von den Möglichkeiten der ihnen zur Verfügung stehenden Weiterbildung und ihrer diesbezüglichen Rechte wissen. Parallel zu den Verpflichtungen seitens der Industrie, der Arbeitgeber und der Regierung soll deren Unterstützung zur Weiterbildung gefördert werden.

8. Die Ingenieurvereinigungen und ähnliche Organisationen spielen eine wichtige Rolle zur Sammlung und Verteilung von Informationen in Bezug auf die Ansprüche der Ingenieure auf Weiterbildung, durch den Kontakt mit Arbeitgebern, Industrie, individuellen Ingenieuren und Lehrinstituten.

9. Alle, die mit dem Lehrberuf befasst sind, sollen dafür sorgen, dass Inhalt und Methoden den Bedürfnissen, Interessen und dem Hintergrund der betroffenen Ingenieure angepasst sind und dass die Ingenieure selbst die notwendige Basis und die Kontakte haben, um ihre Funktion gewissenhaft auszuüben.

## MITTEL FUR DIE WEITERBILDUNG

10. Systematische Studien sollen über den Lernprozess gemacht werden, der die Weiterbildung in sich schliesst. Diese Arbeit soll mit Hilfe der in diesen Disziplinen arbeitenden Menschen ausgeführt werden, wie z.B. Psychologen und Pädagogen. Ausserdem wird empfohlen, dass Organisationen wie FEANI und Unesco das Herstellen von Verbindungen zwischen diesen in verschiedenen Ländern erfolgenden Tätigkeiten fördern sollen.

11. Ein wirklicher Erfahrungsaustausch über Lehrmethoden und Informationssysteme, die für die Weiterbildung zur Verfügung stehen, besonders die neue Methodologie betreffend, soll sich nun abwickeln. Erfolge und Misserfolge von verschiedenen Versuchen sollen von den entsprechenden Befurfsorganisationen in bretem Ausmasse veröffentlicht werden.

12. Nationale Körperschaften, die sich mit Ingenieuren befassen, sollen passende Mittel finden, so dass alle, die sich mit der Weiterbildung der Ingenieure beschäftigen, Industrie und Universitäten miteinbezogen, von den Vorteilen und Beschränkungen, der für die Weiterbildung zur Verfügung stehenden Methoden, unterrichtet sind.

13. Durch entsprechende Gesetzgebung und Finanzaktionen sollen finanzielle Mittel für die Weiterbildung der Ingenieure zugängig sein. Die Verantwortlichkeit des Arbeitsgebers muss auf diesem Gebiet klar festgelegt werden.

# Annexes

# 1. Addresses delivered at the Opening Session

*Opening address by Dipl. Ing. H. Hongisto, Chairman of the Finnish National Committee of FEANI*

Mr Minister of Education, Mr President, Mr Director, Ladies and Gentlemen,

One of the essential functions of FEANI, which now has begun its third decade, is no doubt the one in the field of technical education. Indeed, our annual 'formation' seminars are already becoming traditional.

Some of us may recall the words of Professor Throkil Kristensen at the Århus Seminar in the spring of 1971: 'I think that one of the largest, perhaps the largest, reserve of production capacity in the world today is the imperfect utilization of existing knowledge.' And a little later in the same speech: 'In order to make realities out of the many possibilities created by the new knowledge you must have qualified manpower, and you must have physical capital.'

All of us can probably agree with this thesis that there is a genuine need for our ability to utilize both new knowledge and existing capital to realize our productive potential. To us as engineers this means the creation of better engineers and, in keeping with today's theme, the improvement of existing engineers.

When agreeing with this view we also understand that advancement in the education and training of engineers has become an essential function not only of FEANI, its parallel regional organizations, but has also become an integral part of the programming of organizations engaged in education in general. Here I want to mention in particular the contribution of the United Nations Educational, Scientific and Cultural Organization, Unesco, in the organization of our seminar.

This seminar is the fourth of a series of meetings on Education and Training of Engineers that have been arranged by FEANI since 1969. On this occasion, Unesco is lending its support and we have with us a number of Unesco-sponsored participants from non-FEANI countries. Some other observers have also been invited. We Finns are particularly pleased that the organization of this joint FEANI/Unesco seminar has been entrusted to us. We sincerely hope that the work of the local organizing committee will prove successful and satisfactory to every one of the seminar participants. I wish to thank Unesco for its assistance in the form of expert knowledge as well as its economic contribution to this seminar. I also wish to extend my thanks to the Ministry of Education in Finland as well as to those technical and economic foundations which have given most valuable economic contributions.

I shall not deal with the theme of our seminar, the continuing education of engineers, because two introductory lectures on this subject are included in our programme. My pleasant task is to welcome the participants and lectures of the seminar to Finland. We, the inhabitants of this small northern country, are facing a continuous change in the structure of our economy, for rapid industrialization is still going on. Nearly 150 years ago, our national poet Johan Ludvig Runeberg wrote: 'Our land is poor for he who covets gold', but this is no longer true. We can no longer call this country poor. In terms of gross national income, it is among the fifteen richest countries in the world. And as to gold, this happens to be a by-product of Outokumpu, one of the largest copper mines in Europe. Finland is, of course, still a country of thousands of lakes. Unfortunately, quite a number of them are more or less polluted.

The engineers have played an essential role in the continuously changing society of this country, both in a positive sense but regrettably also in a negative respect, when they have neglected the environment. However, it was stated recently that technology may be the cause but it is also the cure. We engineers certainly have a lot to do in creating the new remedial technology, which is now urgently needed. Especially we Finnish engineers and architects should be the supporters of the creed – if I may so call it – uttered by Jonathan Swift some 250 years ago: '. . . whoever makes two blades of grass grow where only one grew before, deserves better of mankind than any speculative philosopher or metaphysical system builder.'

At the previous seminars of FEANI practically everything has been said about the nature of the education of engineers. Now I would like to emphasize only one feature that was brought forward in Århus last year. All the engineering organizations of the FEANI countries laid stress on the work of engineers and architects, including the education of engineers, as activities in 'the service of humanity'. As an example of this I would like to mention the statutes of the Swedish Association of Engineers and Architects. According to these statutes: '. . . the primary aim of the association is to work for the development of the science of engineering in the service of humanity and the nation, promoting speedy balanced progress, which has regard for man and his environment.'

Among the younger generation in some countries we have already noted a clear decline in interest in the study of technology. We have, at least to some extent, ourselves to blame for this. However, if we succeed in acting 'in the service of humanity', we might also win the confidence and interest of the young. In closing the last FEANI education seminar in London, Professor J. K. Nash, of King's College, London, made the following remark which expresses the view I also hold: 'Engineering is a creative profession. The work of educating engineers is potentially exciting and for the good of mankind: if it is to be made so, there will be a ready partner in the youth of today.'

Ladies and Gentlemen, on behalf of the Finnish National Committee of FEANI, I declare the 1972 FEANI/Unesco Seminar on Continuing Education of Engineers open, and I wish you a hearty welcome to Finland.

### *Welcoming address by Ulf Sundqvist, Finnish Minister of Education*

Mr President, Mr Director, Ladies and Gentlemen,

Education in its many forms takes an increasing share of the efforts of society. Such questions as organizational reforms, public financing of education, the provision of sufficient numbers of trained teachers, equality of educational opportunities, educational research and development, and many other topics are very prominent in public discussions in all countries, and at the international level.

In this country, the particularly large-scale effort we are now engaged in is aimed at the implementation of a nine-year comprehensive school throughout the country, between 1972 and 1977, and the expansion and improvement of higher education, which started in 1966, and for which we have a special development legislation and plans, extending up to 1981. A good many other tasks are either being implemented or are in the planning stages and we could say that education is at present on the move, and is perhaps one of the most dynamic aspects of Finnish society.

The life long education of the working population has recently come into discussion in many countries. This problem is many-faceted and includes several partial problems for the national education systems. One particular field is of course the continuing education of highly trained professional people, whose need to keep up with the general development of their particular field and with their individual changing working responsibilities is obvious.

Some years ago, I read a pamphlet about the problems of re-education and continuing education; to illustrate the situation in engineering, the writer took an example from the United States, where there had been a discussion about the situation in engineering education, and he stated that an engineering chemist five years after he has passed his examination is not only useless but could even be dangerous for society. Perhaps this is going a little too far, but I think it puts the problem in a nutshell.

Although the situation is not exactly catastrophic, the topic of this seminar is a very important one for all societies, for the developed as well as the developing societies of the world, and to paraphrase the speech just made by the Chairman of the Finnish National Committee for FEANI, the cause is of course education, but so also is the cure. Problems are visible in the problems of the present education, but they can be cured only by more education, by better education, and by better-planned education.

Engineers have been almost the pioneers in continuing professional education. Being practical people, they have taken the initiative and proceeded with practical measures through their own organizations in many countries and at the international level. International seminars dedicated the continuing education problems of a particular professional group are, I believe, not common, but it is interesting to learn that this seminar concerning the education of engineers is already the fourth of a series and I think you could give inspiration to other professional groups.

The Government of Finland is very happy to note that Unesco has joined its efforts with those of FEANI in connexion with this seminar. I think that this area of continuing education which has already been discussed at Unesco in many connexions will be, perhaps, one of the most important working fields of this Organization in the future. I am also pleased to note that among the participants there are representatives from other parts of the world besides Europe. We in Europe, and in Finland, too often stick to our own continent and talk about our own problems: the problems that you are about to discuss here are, however, universal.

Ladies and Gentlemen, on behalf of the Government of Finland, I welcome you to our country, and wish every success to your seminar.

### *Welcoming address by the Representative of the Director-General of Unesco*

Mr Minister, Mr President, Ladies and Gentlemen,

May I first of all express, on behalf of the Director-General of Unesco, our gratitude to the Government of Finland for its generous hospitality to this meeting, and in particular our thanks to the Ministry of Education. It is particularly pleasing to be meeting here in a technical university of such renown, and in this regard we are grateful to the authorities of the Helsinki University of Technology.

As you know, this seminar has been organized jointly by Unesco and FEANI. Unesco has arrangements for co-operation and consultation with many international non-governmental organizations conducting activities in fields that are within its competence, and in the engineering field FEANI is one of the several organizations that Unesco has been in contact with over a number of years. The relationship between us has recently grown stronger and we hope it will continue to strengthen in the future, especially with regard to the education and training of engineers.

In this seminar our joint endeavour should result not only in an enrichment of the content of the seminar, but will also mean that its results will be widely disseminated, not only in Europe by FEANI, but also throughout the rest of the world.

I would like to refer briefly to the 1968 Unesco International Conference on the Trends in the Teaching and Training of Engineers, which some of those present may have attended. One of the major recommendations was that more effective programmes of continuing education are called for.

Perhaps even more significant was the recommendation that: 'The initial education of an engineer should be based on the assumption that it will be followed by continuing life-long education, and all those concerned with the engineering profession should plan on this basis.' It is difficult to argue against this latter recommendation, but is it in fact being implemented?

It is appropriate that the first meeting of this type should be held in Europe, for in Europe the concept of continuing education and the practical and legal measures for its implementation have gained more ground than in most other areas of the world. Thus the lessons learned from the European experience may serve to guide countries which are not yet deeply involved in continuing education programmes. This is one of the main roles the present seminar can play, and I hope that you will keep it in mind in formulating your conclusions and recommendations. It appears especially important that there be clear recognition that, because of differences in social systems, educational structures, conditions of employment, or professional bodies, it will rarely be possible for one country to copy the system of continuing education of another; but we can all gain from each other if we share our experience openly.

I hope that during this seminar we will be able to look ahead, even to the year 2000 and beyond. The engineering students of today will then be in responsible positions. What contribution will continuing education have made to their personal development? What means will have been use? Was their initial formal education an appropriate base for subsequent continuing education? These are some of the questions we will be studying in these four days.

Ladies and Gentlemen, on behalf of the Director-General of Unesco, I welcome you to this seminar and wish you every success in your work.

### *Welcoming address by the President of FEANI*

Mr Minister, Mr Chairman, Ladies and Gentlemen,

On behalf of FEANI it is a pleasure for me to greet the delegates to this joint seminar.

First, I would like to thank the Chairman of the Finnish National Committee for FEANI, Mr Hongisto, for his words of welcome. In an international organization like FEANI, all the work which is worth while depends entirely on the efforts of the national members. We are grateful for the work that the Finnish National Committee has done in arranging this seminar, which has made it possible for us to meet and exchange our experiences.

Furthermore, I would like to express our gratitude to Unesco for its invaluable contribution to the joint arrangements. Thanks to the help of Unesco, the geographical borders of the seminar have been extended beyond Europe, which is FEANI's field of work, to include the whole world. I have seen the papers which will be presented here, and it is clear that the geographical expansion has added a great deal to the value of the seminar. Many of the papers report unique experiences and will be very interesting for us here in Europe.

Those present who do not belong to the FEANI circle might be interested to hear some words about the association and its field of activity. Today, the association counts eighteen national societies in the European countries as its members, and the total membership of these societies is between 700,000 and 800,000 engineers.

FEANI's task is to strengthen international co-operation between the national societies of engineers, to work for the status of engineers in society, and especially to study the problems connected with the education and training of engineers.

For this last purpose, FEANI has arranged each year a conference which has dealt with a specific aspect of the education and training of engineers; in 1970 the conference was in Prague, in 1971 it was in London. Here in Helsinki we are going to deal with the continuing education of engineers, and this is a natural continuation of the work.

It might be asked why the national societies of engineers, and through them our international association, have such strong interests in the education and training of engineers. The answer is very simple, namely that education lays the foundation for the future membership of the profession. It is in our interest that our new members should receive an education and training which is as appropriate as possible.

The interest of FEANI's national members embraces both education at the universities and the later training of engineers. In the field of continuing education, we find in many countries that the societies have played an active role by themselves arranging courses. Some of our member societies have large set-ups for this, which arrange a considerable number of courses every year for their members. At first sight you might think that this is not a natural task for the societies, and I think that many of the societies who are engaged in such educational activities might agree. However, in many countries the universities have not had sufficient capacity to take care of this task in addition to initial education. And when none of the universities and schools have been able to cope with the demand, the societies have not seen any other solution than to arrange the courses themselves. I believe, however, that most of the societies concerned wish that the universities could take over large parts of the activities in this field and expand them considerably more than the societies themselves could do.

After these words about FEANI and its connexion with the problems of education, I would like to offer my best wishes to all delegates and also wish every success to the seminar. I am confident that it will produce excellent results.

# 2. Summaries of additional papers submitted to the seminar

*La S.E.E. et la Formation Continue,* by General V. Bertrand (France)

This paper describes the history of the work in the field of education of the Société des Electriciens, des Electroniciens et des Radio-électroniciens. At present three main modes of action for continuing education are pursued, namely the organization of courses, the publication of brochures, and the preparation of videotape material.

*L'Ingénieur et la Formation,* by Mme A. David (France)

The types of information required by an engineer are described, together with their sources and the problems faced in developing more efficient information services.

*L'Effort Français pour la Formation Continue,* by Michel Ferry (France)

The rapid evolution of the status of continuing education in France is outlined, especially the implications of the law of 16 July 1971 in which the right of the individual to continuing education is established and financial support is assured. The role of existing and newly created bodies is reviewed, notably that of the CNIF.

*Continuing Education of Engineers in Great Britain,* by Professor H. J. King (United Kingdom)

After listing the objectives of continuing education, the results of the 1970 survey by CEI are described. The 'total technology' concept of training is then mentioned, as is the need for more rational information on the future demands for continuing education. Finally, the Open University is described.

*Systems Approach and Engineering Education,* by Professor A. Niemi (Finland)

The writer outlines the importance of systems approaches in the education and work of engineers, and analyses the components of engineering curricula. The place of systems subjects in continuing education is then described.

*A Survey of Continuing Education of Engineers in Greece,* by Professor T. Tassios (Greece)

The way in which the initial education of an engineer diminishes in value with time is noted, and its national implications are analysed. The types and timing of continuing education in Greece are described, as is the work of the responsible bodies, especially that of the Technical Chamber of Greece.

*Continuing Education in Engineering with Special Reference to Chemical Engineering,* by Professor R. Edgeworth Johnson (United Kingdom)

This paper was originally published by the Institution of Chemical Engineers. It describes the results of a comprehensive survey of individuals and educational institutions and makes a number of concrete recommendations concerning the expansion of continuing-education courses, based on co-operation between universities and the engineering society.

*Continuing Education of Technologist and Higher Technicians in Cyprus,* by Professor T. Szentmartony (Hungary)

This paper describes a pilot project for the continuing education of higher technicians carried out at the Higher Technical Institute, Nicosia (Cyprus), by national staff and Unesco staff members at the United Nations Development Programme project at the institute.

*Continuing Education of Engineers in Poland,* by Professor J. Tymowski (Poland)

After describing different objectives for continuing education, the writer outlines the history of its expansion in Poland. He mentions that the 1965 Resolution No. 306 of the Council of Ministers provides that systematic perfectioning of employees may become obligatory. In certain cases, concrete examples are then given of various courses. The 1969 action of the Main Technical Organization (NOT) requiring pedagogical training for its lecturers is also mentioned.

# 3. List of participants

*Argentina*

M. A. Yadarola
Comité d'Enseignement UPADI
Mariano Moreno 435
Cordoba

*Australia*

A. H. Corbett
Institution of Engineers
P.O. Box 29
Campbell, A.C.T. 2601

*Austria*

E. Bocovics
Technische Hochschule Wien
A–1130 Vienna
Gutskowplatz 7/14/4

*Belgium*

H. P. Debruyin
Louvain University
Flemish Engineering Society
180 Groenenborgerlaan
B 2020 Antwerp

*Canada*

J. M. Ham
University of Toronto
Room 170
Galbraith Building
Toronto 5, Ontario

*Czechoslovakia*

S. Novak
Technical University
Na Florenci 25
Prague 1

*Denmark*

H. Harboe
Ingeniør Sammenslutningen
Ved Stranden 18
DK-1601 Copenhagen K.

*Egypt*

A. A. El-Erian
University of Cairo
Cairo

*Finland*

L. Blomstedt
Postgraduate Education Centre
of the Engineering Societies in Finland
INSKO
Annank 18
00120 Helsinki 12

C. E. Carlson
Suomen Teknillinen Seura
Merikatu 3 B
00140 Helsinki 14

C. E. Hohenthal
Tekniska Föreningen i Finland
Runebergsg. 36 A 6
00260 Helsinki 26

H. Hongisto
Finnish National Committee of FEANI
Suomen Sokeri
02460 Kantvik

J. Ihaumuotila
Valmet Oy
Punanotkonkatu 2
00130 Helsinki 13

J. E. Jansson
Helsinki University of Technology
TKK
02150 Otaniemi

M. Kaario
Suomen Teknillinen Seura
Yrjönkatu 30
00100 Helsinki 10

R. Keskinen
Tampere University of Technology
Rautatienkatu 7 A 9
33100 Tampere 10

E. Kinnunen
Lappeenranta University of Technology
Tullitie 6
53100 Lappeenranta 10

U. Kanttinen
Oulu University
Kihokkitie 12
90120 Oulu 12

P. Laasonen
Helsinki University of Technology
TKK
02150 Otaniemi

B. Myréen
Abo Akademi
Kuppisg. 9 C
20520 Abo 52

A. Niemi
Helsinki University of Technology
TKK
02150 Otaniemi

C. G. Nordman
Insinööriliitto
Tennisstigen 2 E 53
02100 Tapiola

M. Nurmi
Ministry of Education
Rauhankatu 4
00170 Helsinki 17

E. Peippo
Suomen Teknillinen Seura
Yrjönkatu 30
00100 Helsinki 10

V. Porra
Helsinki University of Technology
TKK
02150 Otaniemi

O. Ranta
Ministry of Education
Rauhankatu 4
00170 Helsinki 17

L. Rousi
Finnish National Board
of Vocational Education
Ammattikasvatushallitus
Eteläesplanadi 10
00130 Helsinki 13

A. Saarialho
Helsinki University of Technology
TKK
02150 Otaniemi

J. Salmels
VTT
02150 Otaniemi

E. Törnudd
Helsinki University of Technology
Library
TKK
Kirjasto
02150 Otaniemi

V. Virkkala
Kone Osakeyhtiö
Munkkiniemen kartano
00330 Helsinki 33

J. Virkkunen
Helsinki University of Technology
TKK
02150 Otaniemi

E. Tunkelo
Helsinki University of Technology
TKK
02150 Otaniemi

*France*

V. Bertrand
Société des Electriciens,
des Electroniciens
et des Radio-Electroniciens (SEE)
10 Avenue Pierre Larousse
92 Malakoff

M. T. Monticone
Conseil de l'Europe
67 Strasbourg

M. Pelegrin
École Nationale Supérieure de
l'Aéronautique et de l'Espace
(ENSAE)
B.P. 4032
31055 Toulouse

L. Ter-Davtian
OCDE
Boulevard du Montparnasse
Paris-6$^{e}$

*Federal Republic of Germany*

K. H. Buschmann
Verband Deutscher Elektrotechniker (VDE)
6 Frankfurt/Main 70
Stresemann Allee 21

J. H. Hahn
University of Karlsruhe
Institut für Siedlungswasserwirtschaft
Universität Karlsruhe
75 Karlsruhe 1
Kaiserstr. 12

R. W. Peters
Verein Deutscher Ingenieure (VDI)
4 Düsseldorf 1
Graf-Recke Str. 84

M. Polke
Deutscher Verband Technisch-Wissenschaftlicher Vereine (DVT)
4 Düsseldorf 1
Graf-Recke Str. 84

J. Rink
Verein Deutscher Eisenhüttenleute (VDEh)
4000 Düsseldorf 1
Breite Str. 27
Postfach 8209

*Greece*

L. Logothetis
Université Technique d'Athènes
15 Rue Tossitsa
T.T. 148
Athens 103

*Hungary*

O. Petrik
METESZ
Budapest XI
Müegyetem rkp 3

Imre Szebényi
METESZ
Budapest XI
Budafoki ut 8

*India*

S. K. Bose
Indian Institute of Technology
Kharagpur
West Bengal

*Ireland*

M. D. O'Donnell
College of Technology
11 Ailesbury Lawn
Dundrum
Dublin 14

M. B. O'Donovan
National Institute for Physical Planning and Construction Research
St Martin's House
Waterloo Road
Dublin 4

J. P. Sheehan
City of Dublin Vocational Education Committee
Town Hall
Ballsbridge
Dublin 4

*Israel*

E. Lahav
Government Office of Chief Scientist
Hakiria
Tel Aviv

*Italy*

G. Casalis
Consiglio Nazionale Degli Ingegneri (CNI)
Via Bertoloni, 31
00197 Rome

A. Ferrari-Tonolio
Consiglio Nazionale Ingegneri (CNI)
Via Nicotera, 29
Rome

*Japan*

T. Mizushina
Kyoto University
Dept. of Chemical Engineering
Kyoto University
Kyoto

*Luxembourg*

G. Steichen
Ecole Technique
19 Rue Guillaume Schneider
Luxembourg

*Netherlands*

J. C. Goedkoop
Koninklijk Instituut van Ingenieurs
Netherlands National FEANI
Committee
23 Prinsessegracht
The Hague

J. A. Sorel
NIRIA
11 Nieuwe Parklaan
The Hague

W. A. Vervoort
NIRIA
11 Nieuwe Parklaan
The Hague

J. C. Vlugter
Netherlands National FEANI
Committee
23 Prinsessegracht
The Hague

*New Zealand*

A. L. Titchener
University of Auckland
Private Bag
Auckland 1

*Norway*

A. Hernes
Den Norske Ingeniørforening
NIF
Kronprinsensgt. 17
Oslo 2

I. Holand
Den Norske Ingeniørforening
NIF
Kronprinsensgt. 17
Oslo 2

A. Reistad
Den Norske Ingeniørforening
NIF
Kronprinsensgt. 17
Oslo 2

B. Slungaard
Den Ingeniørforening
NIF
Kronprinsensgt. 17
Oslo 2

*Poland*

J. Tymowski
Polish Central Organisation
Aleje Niepodl. 165 m 8
Warsaw

*Romania*

C. Mocanu
Polytechnic Institute of Bucharest
Bucharest
Romania

*Spain*

M. Jerez Juan
Instituto de Ingenieros
Civiles de España
Ferraz 24
Madrid 8

V. Obach Planas– Dorai
Centro de Perfeccionamiento
del Ingeniero
Vía Layetana 39-40
Barcelona 3

*Sweden*

B. Ando
Svenska Ingenjörssamfundet
Apnäsvägen 40, 10 tr
S-175 37 Järfalla

K. E. G. Gestrelius
Malmö School of Education
Lärarhögskolan
Fack
S-200 45 Malmö 23

B. G. Hagy
Svenska Teknologföreningen
Box 40 116
S-103 43 Stockholm 40

E. B. Sjögren
Svenska Teknologföreningen
Box 40 116
S-103 43 Stockholm 40

*Switzerland*

W. Gysin
Schweizerischer Technischer
Verband STV
Postfach CH-8023
Zurich

J. C. Piguet
FEANI
Ecole Polytechnique Fédérale
Lausanne

P. J. Profos
Technische Hochschule
Institut für Regelung
und Dampfanlagen
Sonneggstr. 3
CH-8006
Zurich

U. Zürcher
Société suisse des ingénieurs
et architectes
B.P. 8039
Zurich

*United Kingdom*

W. L. J. Farvis
University of Edinburgh
14 Cluny Terrace
Edinburgh

H. L. Kings
Council or Engineering Institutions
(CEI)
Mining Department
The University
Nottingham

M. W. Leonard
Council of Engineering Institutions
(CEI)
2 Little Smith Street
Westminster
London SW1P 3OL

J. C. Levy
The City University
18 Woodberry Way
N. Finchley
London, N 12

D. H. Sharp
Institution of Chemical Engineers
16 Belgrave Square
London SW1X 8PT

Sir Frederick Edward Warner
Cremer and Warner
140 Buckingham Palace Road
London SW1W 9SQ

*United States of America*

M. R. Barry
The University of Wisconsin
439 Engineering Research Building
1500 John Drive
Madison, Wisconsin 53705

J. M. Biedenbach
RCA Corporation
Building 2–8
Camden
New Jersey 08102

J. P. Klus
The University of Wisconsin
432 North Lake Street
Madison, Wisconsin 53706

M. W. Krigel
Oklahoma State University
Stillwater
Oklahoma 74074

A. G. Schillinger
Polytechnic Institute of Brooklyn
333 Jay Street
Brooklyn
New York 11201

*U.S.S.R.*

A. Bogomolov
U.S.S.R. Ministry for Special
Higher and Secondary Education
Zdanoda Street 11
Moscow

M. Poltev
U.S.S.R. Ministry for Special
Higher and Secondary Education
Zdanova Street 11
Moscow

*Venezuela*

V. Yackovlev
National Council of Universities
Apartado 51566
Caracas 105

*FEANI*

V. Broida
Secretary General Designate
FEANI
4 Rue de la Mission Marchand
75016 Paris

R. Bureau
General Secretariat
FEANI
4 Rue de la Mission Marchand
75016 Paris

G. Clogenson
Secretary General
FEANI
4 Rue de la Mission Marchand
75016 Paris

O. Irgens
President
Olav Aukrustvei 27
Oslo 3

*Unesco*

A. Evstafiev
Director, Division of Technological Research and Higher Education, SCT
Unesco,
7 Place de Fontenoy
75700 Paris

A. S. Goodyear
Chief, Engineering Education Section
Division of Technological Research
and Higher Education, SCT
Unesco,
7 Place de Fontenoy
75700 Paris

# 4. Members of the local organizing committee

H. J. Hongisto, Chairman of the Finnish National Committee of FEANI
J. E. Jansson, Professor, Chairman of the Formation Committee of FEANI
Matti Kaario, General Secretary, Finnish National Committee of FEANI
Osmo Ranta, Secretary, FEANI Formation Committee
K. G. Löfström, Secretary, Organizing Committee